Strawberries, Peas & Beans

Truck Farming in Anne Arundel County

By Willard R. Mumford

ISBN 0-9702355-0-X
Library of Congress Catalog Number: 00-106394

Book design by Betsy Earley of Bay Media, Inc.
Cover design by Eileen Mumford

Project management, consulting and design by Bay Media, Inc.
1244 Ritchie Highway, Suite 6
Arnold, MD 21012-1887
410 647-8402 Fax: 410-544-4640
Web site: http://www.baymed.com

DEDICATION

This book is dedicated to the memory of Violet Mumford and Ora Pumphrey Smith, the two women who were instrumental in collecting the details of this story. They kindled my interests in history and in truck farming, which motivated me to document this fascinating but relatively unknown facet of Anne Arundel history. Violet Mumford, my mother, gave me an intense interest in appreciating my past and served as my contact with the Ann Arrundell County Historical Society while I was in the U.S. Air Force.

Ora Pumphrey Smith, working as a volunteer for the Ann Arrundell County Historical Society, became known as the Pickers' Check Queen while uncovering thousands of the little brass tokens known as picker checks. Ora grew up on a truck farm and promoted the memory of the truck farming days. The jewelry made from picker checks by Mrs. Smith raised over $60,000 for the Historical Society. She introduced me to the world of truck farming in old Anne Arundel.

Violet M. Mumford

Ora Pumphrey Smith

PREFACE

I left Oregon when I had just turned nine years old but one of my fond memories was during the summer when we went "picking." Any kid could do it, and crowds of youngsters would gather at the city park early each morning. There, farmers would haul us to their fields. It was an adventure! We never knew from day to day what we were going to be picking. I remember apples, plums, peaches, hops, and strawberries. Strawberries were my favorite...probably because they tasted so good. I'm sure my mouth was as stained as much as my hands when I turned in my filled boxes for chits. Little did I realize that years later and over 3000 miles away I would again encounter a strawberry culture which had long since disappeared, but in a very real sense was related to my heritage. This culture was a part of the market gardening, or truck farming, of Anne Arundel County, Maryland. This time, however, the stain was in my heart.

Truck farming was an institution that was to last for over a hundred years. Nobody knows exactly how or when it started. Truck farming was a natural for Anne Arundel County with its loamy sand soil and easy access to the rapidly growing urban area of Baltimore. Little brass picker checks are about all that is left of the truck farming culture. It was the tokens that caught my eye.

I first encountered the tokens at an antique flea market in the summer of 1969. The shiny brass tokens in a dish on one of the tables caught my numismatic eye. I picked up several of the tokens and asked the lady behind the table, "Do you know what these were used for?" She wasn't sure, but thought they were local farm tokens. I asked more questions, and it became obvious to the lady that my curiosity was more than just a passing fancy. "You can find out more about them at the Ann Arrundell County Historical Society's Browse and Buy Shoppe in Arnold. Ask for Ora Smith. She's the one who knows all about them."

At the very next opportunity, I drove to the old railroad power house near Jones Station, which had been converted into an antique shop by the Ann Arrundell County Historical Society. It was there that I first met Ora Smith. She was delighted to introduce me to Anne Arundel County pickers' checks. A fascinating story unfolded. Ora Smith had grown up on a truck farm and had lived with strawberries, peas, beans, migrant workers and picker checks. I was intrigued, but did not then fully realize the extent of the story.

Once again, as is so often the case, a numismatic treasure has opened the door to another world. This time the new world was in my own backyard and was indeed an interesting part of our Anne Arundel heritage. I also had a feeling that I would like to be involved in the preservation if not the resurrection of that world.

I enjoyed my visit with Mrs. Smith, which included more pickers' checks and the opportunity to hang curtains in all the windows of the little shop. I left with a warm feeling in my heart, a little sweat on my brow...and a membership card in the Historical Society.

At every opportunity I gathered as much information as I could about picker checks and the system which gave them birth. This was a whole new world to me, not only in numismatics, but also in a neglected period of Maryland history...a period which spanned a unique cultural experience lasting over 100 years. It is not found in the history books and is a mixture of facts and legends, sprinkled with the dust of dreams. The story has at least four interesting facets. There were the farms and the farmers, the migrant workers or "pickers" who came to live on the farms during the picking season, the produce, and the little odd shaped tokens, bearing simply initials and numbers. Together the farmers, the produce, the pickers, and their checks constitute a period of history in Anne Arundel County which preceded its maturity as an urbanized, recreational bedroom community for the Washington-Baltimore complex.

Through the years, working with Ora Smith and my mother, Violet Mumford, my collection of picker checks grew beyond my greatest expectations. As I was not living in the area, my mother coordinated my collection with Mrs. Smith. Whenever Ora would locate new tokens, she would contact my mother who would purchase several sets for me. Mom would carefully record and sketch each token for her own files and then send the tokens to me. Every birthday and Christmas I would receive a package which had the familiar jingle of picker checks. It was always the highlight of the occasion, and I immediately and joyously added the new checks to my inventory. I have been amazed at the number of tokens that have surfaced over the years.

In 1970, I had tokens from 32 different farmers. Four years later the list grew to 72 farmers with 433 different varieties. After I moved back to Maryland in 1979, my list grew rapidly. I started taking inventories of the largest known collections, and the list now exceeds over 400 farmers and 1500 different varieties. Many of the tokens are unidentified, as there is no record of the farmer who used them. Compounding the problem of identifying the tokens is the existence of a large number of packing or canning tokens which look very much like the picker checks. Many of the canneries operated in the Baltimore area, but most of the canning tokens come from the canneries on the Eastern Shore of Maryland. Only recently have the canning tokens been cataloged and these are not listed in this volume. There are still over a hundred unidentified farmer's initials found on tokens. In many cases these are known to exist only as a single token. Some cannot be documented as authentic farm tokens, but those that are identified as tokens from Anne Arundel County are listed in this catalog. Many families of farmers remember that their father or grandfather used tokens, but don't know what became of them. The farmers are listed even though no tokens are now known to exist.

Another problem arises when one tries to limit the picker checks to the geographical boundaries of Anne Arundel County as we know them today. While the great concentration of farms which used the tokens was located in northern Anne Arundel County, the use of picker checks was by no means limited solely to this area. Although picker checks were used in many counties in Maryland and similar tokens were used in other states, there has never been as great a concentration of farmers using picker checks as in Anne Arundel County.

Every year since 1969 more tokens have been discovered and, no doubt, many more will surface in the years to come. Perhaps someday a supplement will be necessary to update current inventories of picker checks. But this story remains a very important and largely neglected part of our Anne Arundel heritage.

So many individuals have contributed to the research effort which has made this document possible. I would like particularly to thank Bill Kuethe, Ben Egerton, Emma Schramm, John Prechazka, Jim Boswell and the late John Henry, William Sanders and Vernon Taylor for allowing me to inventory their collections. I am also very grateful to Isabel Cunningham, John Grabowski, the late Mabel Parker, John Mellin, the Ann Arrundell County Historical Society, Maryland Tokens and Medals Society, Roger White, Jack Kelbaugh, John Smith and his late wife Ora Smith, Jim and Mary Calvert, Jimmy Schillinger, Sara Anne Stinchcomb, Margaret Rose, Dolores Klingelhoefer, Rachel Hall Brown, and Jim Boswell for information. Special thanks to Isabel Shipley Cunningham for sharing her farm heritage and her generous and careful assistance in editing and to Jim Boswell who has gone house to house, cemetery to cemetery, seeking information on pickers' checks and the farmers and has assisted me in compiling the inventory of pickers' checks. I would also like to thank Mark Schatz, vice president of publications of The Ann Arrundell County Historical Society for his assistance in making this project possible.

Without these people and many more, this volume would never have been published.

Bracelet made from pickers' checks.

STRAWBERRIES, PEAS, AND BEANS
Truck Farming in Anne Arundel County

TABLE OF CONTENTS

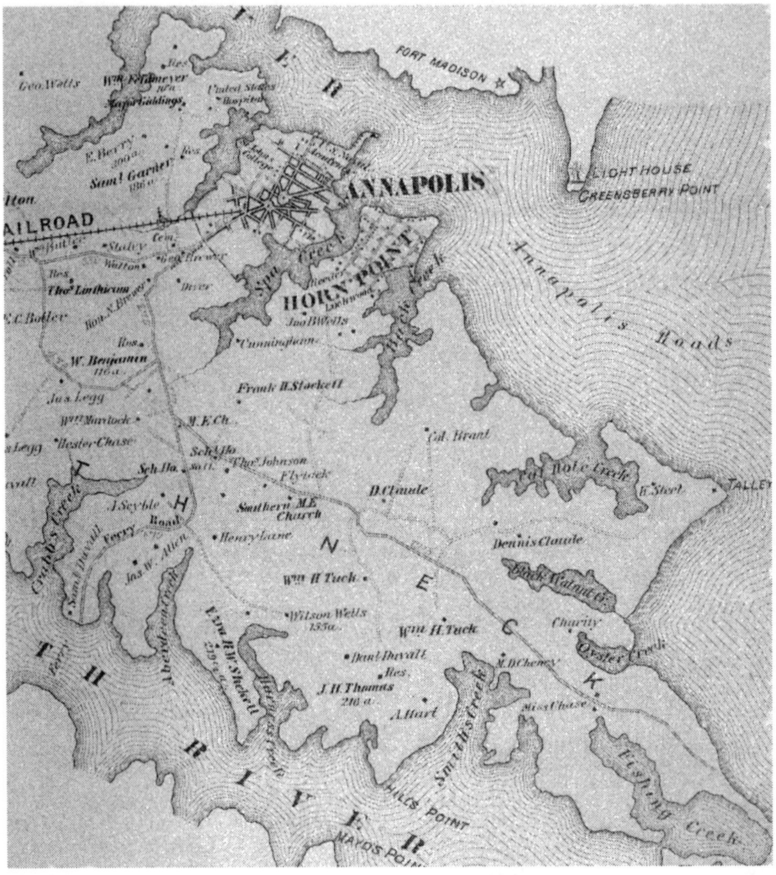

The Early Years

CHAPTER ONE: TRUCK FARMING - THE EARLY YEARS

The year 1649 is one of the more significant years in history. Oliver Cromwell won his civil war against the English crown and established the Commonwealth; Charles I was executed, and the beloved wife of Lord Baltimore II, Anne Arundel, died in her 34th year. In the same year, a small group of Puritans fleeing religious persecution in Virginia settled on a peninsula of land separating the Severn River from the Chesapeake Bay in Maryland. Within a year, the settlement on Greenbury Point, called Providence, was large enough to be officially recognized by the Maryland General Assembly which declared,

> "that part of the Providence of Maryland, on the west side of the Bay of Chessopeack, over against the Isle of Kent, formerlly called by the name of Providence by the inhabitants there residing and inhabiting this yeare, shall henceforth be created into a shire, or county, by the name of Annarundell county."

Following St. Mary's and Kent Counties, Anne Arundel was the third county to be settled in Maryland. The boundaries were not clearly defined and its present configuration was not finalized until two hundred and seventy years later. In 1659, the northern half of Anne Arundel became Baltimore County; in 1851 Anne Arundel was again split in the northwest to form Howard County, and in 1919 the northern boundary retreated south as part of Anne Arundel County was annexed by the City of Baltimore.

Anne Arundel was left with 419 square miles of an interesting mixture of fertile land bordered by Baltimore City on the north, the Chesapeake Bay on the east, Calvert County on the South and the Patuxent River on the west. With the Patapsco, Magothy, Severn, South and West Rivers cutting deep into the county from the Chesapeake, the county claims over 400 miles of shoreline. The land adjacent to the waterways is moderately hilly and was originally covered with a dense forest of oak, maple, tulip poplar, chestnut, pine and a wide assortment of smaller trees. Inland the land is more level, but maintains a rolling countryside.[1]

Anne Arundel has a wide variety of soils left by deposits from the sea, which once covered the area. The Sassafras, Collington, and Norfolk soils are the most important agricultural soils in the county, and in varying forms, lend themselves ideally to support an agricultural economy based on tobacco, wheat, corn and a wide variety of vegetables and fruits. The fertile soil with a temperate climate, abundance of game and seafood, and accessibility to shipping would prove a boon to the early settlers of the county.[2]

Just as in Virginia and in St. Mary's County, Maryland, the early settlers of Anne Arundel County planted corn and a wide variety of other vegetables for their subsistence and tobacco for trade. The economy of early Maryland was based primarily on tobacco, as the "sotweed" not only provided a livelihood for the colonists, but literally served as their money for over a hundred years. Many of the 18th century colonial homes found in Annapolis and throughout the county were built from the profits of tobacco. Salaries, fines, and taxes were paid in pounds of tobacco. Practically every colonist who owned land grew tobacco. But tobacco was hard on the fertile soil. After a few years of production, the soil was depleted of the necessary nutrients to support quality tobacco. Over-production also created a greater supply than could be marketed. These factors forced many farmers to diversify their crops by growing wheat and other grains.

The capital of Maryland was moved from St. Mary's City to more centrally located Annapolis in 1694. The new capital became one of the leading cities of colonial America and reached its zenith just prior to the American Revolution. After independence was achieved, Baltimore, with its better port facilities, soon became the leading city of Maryland, leaving Annapolis in a cultural and economic depression.

The Industrial Revolution stimulated Baltimore's growth while a great transition occurred in Anne Arundel County. With the soil badly depleted from years of growing tobacco, Anne Arundel farmers depended even more on wheat, corn, small fruits and vegetables. The rapidly growing Baltimore population provided a ready market for the vegetables and fruits grown in Anne Arundel. The construction of the Elk Ridge and Annapolis Railroad that connected Baltimore and Annapolis in 1840 provided rapid transportation to the Baltimore markets and opened new lands to farmers who now had the waterways on the east and the railroads on the west. Market access proved the catalyst for a boom in truck farming.

From the 1840s through the next hundred years, truck farming grew to become the dominant type of agriculture in northern Anne Arundel County. Farmers started growing vegetables and small fruits to supply the markets of Baltimore, and later, Anne Arundel produce was shipped as far north as Montreal and as far west as Denver. The northern section of the county was evolving into one of the great trucking areas of America and was especially noted for its quality produce. Already by 1850, Anne Arundel County had become one of the leading strawberry districts in the country.[3]

In the southern half of the county tobacco remained king. Some farmers living close to the waterways converted part of their land to truck produce, and many maintained fruit orchards, but their major cash crop was still tobacco.

The United States Naval Academy was established in Annapolis in 1845 by the Secretary of the Navy, George Bancroft. The capital city experienced some stimulation as its economy became largely dependent upon the growth of the Naval Academy.

No battles were waged in Anne Arundel County during the American Civil War, but the county played a vital role in logistics. Because of heated southern sentiments in Baltimore, President Lincoln agreed not to send any Union troops through the city. Troops were transported by rail to Havre de Grace, just north of Baltimore, and then by ship to Annapolis. There they would board the train for Washington, D.C. and points south. The railroad from Annapolis remained a strategic line for troops and supplies throughout the war. Union troops were stationed along the tracks to protect them from sabotage by Southern sympathizers.

A large Civil War camp, about a mile west of Annapolis, was used as a recovery and retraining area for Union soldiers who had just been exchanged or paroled from Southern prisons; hence the name Camp Parole. With the large influx of soldiers into the county, and to help feed the large Union army, the demand for food increased. Farmers were encouraged to produce more vegetables and other items of food. Truck farming took a giant step forward as a result of the war years and rapidly gained preeminence in the northern part of the county.

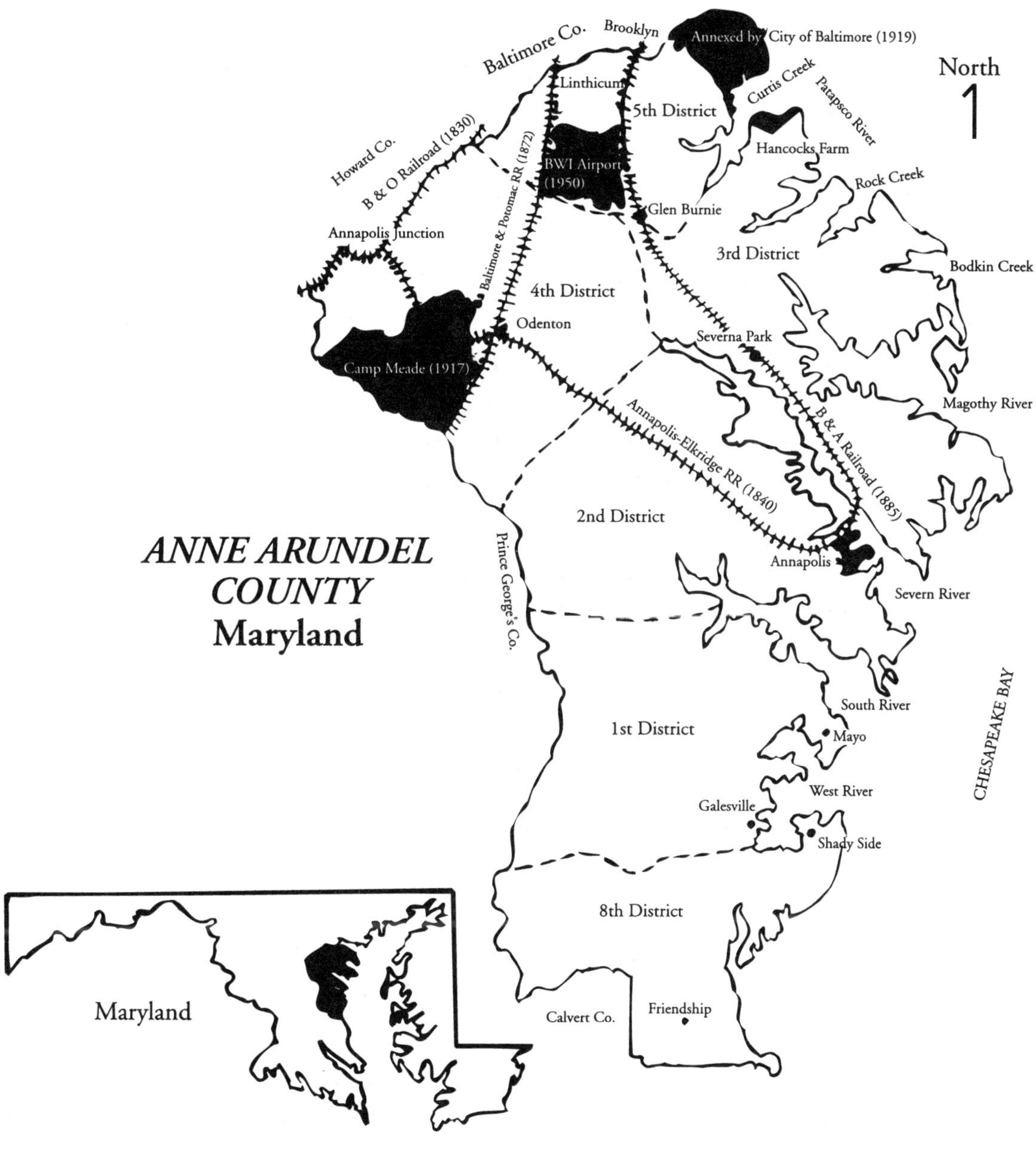

The Growth of Truck Farming

Anne Arundel farmers responded to the great need for more farm produce created by the Civil War by increasing their production of vegetables and fruits. Baltimore had become a canning center for oysters as early as 1849, and it was a matter of logic to add fruits and vegetables to the canning process. The oyster season started in the fall and peaked in the winter months, while the truck produce started in the spring and concluded by the end of the summer. The canneries could remain open the entire year with the addition of fruits and vegetables. This increase in demand, along with the rapidly growing population of the city of Baltimore and the eventual construction of more railroads, led to an even greater growth in truck farming in Anne Arundel County. The labor shortage for the picking season was solved by bringing thousands of immigrants living in Baltimore to the farms for the summer.

Truck farming flourished in Anne Arundel County until the late 1920s and early '30s, when refrigerated trucks brought produce from the South to the Baltimore markets before the Anne Arundel produce was ripe. The great expansion of the Baltimore-Washington area, triggered by the automobile, also caused land prices to increase so much that many farms were sold for developments. The farms have since turned into highways, residential areas, shopping centers and industrial parks. Very few of the old farms remain in the northern half of the county. One of the last remaining relics of the old truck farms are the little brass tokens which the farmers used to pay their pickers.

The county has now become a bedroom community for the residents of the Baltimore-Washington metropolitan area, and is a major recreation area for boaters with well over a hundred commercial marinas and thousands of boats. Annapolis is once again a major port and claims to be the sailing capital of America.

Anne Arundel County remains a land of historic charm and great natural beauty. Citizens who live here are proud of their rich heritage and concerned about its future.

Population x 1000
Baltimore City and Anne Arundel County
1790 - 1930

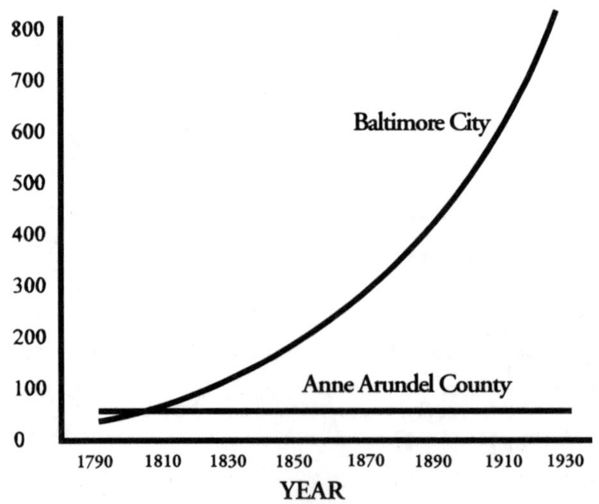

The System

Definitions:

Truck Farming - "The growing of vegetables for a commercial market." Farming in Anne Arundel County would be more accurately described as "market gardening," as the farmers grew a diverse quantity of produce and depended largely upon their own resources to transport the produce to the markets. Today truck farms are much larger and specialize in one or several products which are shipped to markets all over the world. In addition to vegetables, most Anne Arundel farms also grew a variety of fruits for market, but contemporary references refer to the Anne Arundel farms as truck farms; hence for this document the term "truck farm" will be used for farms growing both vegetables and fruits.

Truck produce - Produce grown on a truck farm, i.e. vegetables, and in Anne Arundel County, this term includes a wide variety of fruits.

Truck - The term "truck" as used in the days of the old truck farms is often confused with the term "truck" as used today to denote a motored vehicle used for hauling materials. The Universal Dictionary of the English Language of 1898 defines "truck" as "Commodities suitable for barter or small trade, hence, small commodities; specif., in the United States, vegetables raised for market." "Truck" as used on the old farms is synonymous with the word "produce," which on Anne Arundel farms consisted of vegetables and fruit. This word comes from the Old French, *troquer*, meaning "small articles or vegetables." By coincidence, the large wagon used to transport truck produce to market, when motorized, became known as a truck, which is derived from the Greek, *trochos*, meaning "wheel or a wheeled vehicle." They are two different words which came together as motorized trucks were used to haul truck (produce) to the markets.[4]

Since early colonial times, Anne Arundel County, Maryland, outside of the capital of Annapolis, was an agricultural community largely dependent upon the growing of tobacco. In the early days tobacco was the Marylanders' "meat, drink, clothing and money."[5] The economy was so dependent upon tobacco that farmers and gardeners produced as much of the leafy weed as ground permitted. The soil in the southern part of the county was better suited for tobacco, and most of the large plantations were located south of the Severn River.

The few colonial farmers in the northern part of the county also grew tobacco, but soon found that other produce was better suited to the soil. With the development of several iron furnaces, large sections of land were cleared of their trees to provide fuel for the furnaces. Due to this stripping of land for farms and furnaces, by 1750 over half the forests in the county were gone. With the great population increases after the Revolution, much of the already cleared land in the northeastern section was put to good use in growing vegetables and fruits, which could easily be shipped by water to the Baltimore or Annapolis markets.

As early as 1815, perhaps sooner, farmers in Anne Arundel County began growing strawberries for the Baltimore markets, laying the foundations of an industry which gave the county preeminence for many years.[6] Along with the berries, vegetables of many kinds flourished in the sandy loam of Anne Arundel County. The Elk Ridge and Annapolis Railroad, built in 1840, brought the Baltimore and Washington markets closer to the inland farms, and with the influx of thousands of Union troops during the Civil War, the demand for food soared. More farmers turned to the production of vegetables and fruits as prices also soared.

After the war, with the departure of the troops, a general depression fell over the county. Farmers lost a major source of cheap labor as the slaves were freed. Many former slaves stayed on the farms with the farmers, but more fled the country for the cities where better jobs could be secured. Farm prices fell and the labor shortages left the farmers helpless. But the Baltimore population continued its fast growth and farmers soon found the markets demanding more truck produce. Labor problems continued to plague the farmers, especially during the peak period when the fruits and vegetables needed to be harvested. In the early 1870s farmers found they could recruit immigrants from Baltimore to pick the produce.

The expansion of the oyster canning industry to include fruits and vegetables created an even greater demand for truck produce. In 1872, the Baltimore and Potomac Railroad opened a line running through the northwestern part of the county, which provided rapid market access for more of the farmers in that section of the county. With the Annapolis and Elk Ridge Railroad running along the south side of the Severn River, northwest to Annapolis Junction, and steamers making daily runs into Baltimore from the shore areas, truck farming flourished. Gradually, more farmers turned to the very profitable trucking, drawing more immigrant pickers from their Baltimore homes. With over 4,000 farms in Anne Arundel County, truck farming reached its peak around the turn of the century. More than 10,000 pickers migrated from Baltimore to live on the Anne Arundel farms for six to eight weeks, picking strawberries, peas, beans...and other produce. Living in shanties or barns, the pickers created their own legends, becoming a colorful part of the truck farming culture.[7]

Anne Arundel County, known for its quality produce, supplied a large percentage of the fruits and vegetables for the markets of Baltimore, Washington, and Annapolis. However, local consumption was small compared with the produce shipped to most of the major cities in the northeast as far as Montreal and as far west as Denver.[8] Strawberries, peas, beans, corn, sweet potatoes, asparagus, berries, melons, cantaloupes, and numerous other produce carried a premium when tagged with an Anne Arundel label.

The farms and the farmers formed the nucleus of the truck farming culture. Experiencing the Industrial Revolution, the farms evolved from almost pure hand labor to a highly mechanized system which increased the efficiency of the operation. Electricity, the steam engine, and later the gasoline engine revolutionized the farm as well as the rest of society. The dedication of farmers has always fed the world, but the story of the Anne Arundel truck farmers is more than just a page from our county's history; it is a part of our culture and heritage. Anne Arundel truck farming has been neglected in Maryland history books, and yet, it is the story of our ancestors. It is significant in the formation and growth of our present society.

The final facet of this story is minor, but important, as it is one of the last remnants of the truck farming culture: the small brass and aluminum tokens which the farmers used to pay their pickers. Pickers' checks, with the farmer's initials and a value stamped on one side, are not exciting numismatically, but are historically exciting. Made for function and not for advertising as many tokens are, these checks remain relics which help to document the extent of truck farming in the county.

Truck farming in Anne Arundel County flourished through the first three decades of the Twentieth Century, but with the advent of refrigerated trucks and railroad cars, produce from the Carolinas and Georgia, where the growing season is earlier, reached Eastern markets long before the Anne Arundel produce could ripen on the vine. The availability of early produce from the South drove the prices down for the later local produce and undermined much of the profit of local farmers. Throughout much of the late twenties, Anne Arundel production decreased. With the rising costs of land for industry and housing developments, many farmers sold their farms.

The affordable automobile also contributed to the demise of the Anne Arundel farm as middle class Americans could now own their own automobiles. This permitted them to live greater distances from their working place and initiated the growth of the urban sprawl. More and more Americans were buying former farmland in the country to escape the busy, crowded life of the city. The automobile also replaced the horse as the main means of urban transportation, which eliminated "street dirt" or manure, which enabled the farmer to keep his fields rich with organic materials.[9] All of these factors contributed to the downfall of the Anne Arundel truck farm, but the final blow was the Great Depression of 1929!

After World War I, many farmers outside Anne Arundel County, especially on the Eastern Shore, started growing more vegetables for urban markets as the great demand and prices for export grain and tobacco dropped drastically. This conversion to truck produce lowered the prices for vegetables and fruits. Farmers were going into debt long before the depression hit and had no chance to recover after the collapse of the economy. By 1933, one-fourth of Maryland farmers were approaching bankruptcy and over 100,000 people in Baltimore were out of work with no income.[10] Farmers continued to grow produce, but with little cash, people could not purchase as much food as before the depression. This was especially true of "luxury" foods like strawberries, tomatoes, peaches, and melons, all of which were major crops in Anne Arundel County. Gross income from crop production in Maryland dropped by over 60% from 1929 to 1932. In 1929, the income from truck crops and fruits was over $27 million. By 1932, this income dwindled to just over $11 million. More farmers went into debt, and many were forced to sell land to avoid bankruptcy.[11]

By the start of World War II, many truck farms had disappeared; however, the demand for food increased and many farmers went back into the business of raising fruits and vegetables. Large developments were put on hold during the war which made more land available for farming. Maryland farmers, though faced with severe labor problems, produced more food to support the war.

After the war, large developments started absorbing farm land and, within a few years, the predominance of the Anne Arundel truck farm was history. With a declining demand from urban markets, several hundred roadside stands sold the produce of local farmers. Some Anne Arundel truck farmers returned to growing tobacco, which soon became their major cash crop. The remnants of truck farms are still found in roadway stands and in several farmer's markets which allow farmers to sell their produce to county residents. But the truck farm that once dominated the northern half of the county is gone forever.

The Ann Arrundell County Historical Society is currently developing its headquarters, the Benson-Hammond House on the northeast corner of the Baltimore-Washington International Airport (BWI), into an Interpretive Truck Farm. The house was built around 1830 and operated as a truck farm for over a hundred years. The society is relocating outbuildings from other farms in the area and will help to preserve the truck farming culture of the county. When this project is completed, a tour of the farm will explain the functions of each facet of the old Anne Arundel truck farm. The culture and the history of this era will forever be preserved for all citizens.

Benson-Hammond House (c1830). Home of truckfarmers Thomas and Joseph Benson and John and Rezin Hammond. The home was purchased by the city of Baltimore in 1947 for Friendship Airport. The house was restored by the Ann Arrundell County Historical Society in 1978 and is now used as their headquarters. An "interpretive truck farm" is now under development. (Photo courtesy of the Ann Arrundell County Historical Society).

CHAPTER ONE NOTES

1. For an excellent brief history of Anne Arundel County, see Chapter 1 of *Anne Arundel County, Maryland - A Bicentennial History, 1649-1977,* written by Jane McWilliams. Also see: *Anne Arundel's Legacy - The Historic Properties of Anne Arundel County* by Donna Ware, a project of the Environmental and Special Projects Division, Office of Planning and Zoning, Anne Arundel County, Maryland. See also **http://www.co.anne-arundel.md.us/aaac_history.htm**
2. For a comprehensive analysis of the soils of Anne Arundel County, see: *Soil Survey of Anne Arundel County, Maryland,* by S.W. Phillips and S.O. Perkins, United States Department of Agriculture and H.B. Winant, Maryland Agricultural Experiment Station; Bureau of Chemistry and Soils; United States Department of Agriculture; Series 1928, Number 18.
3. Fletcher, Stevenson W., *The Strawberry in North America* (New York: The MacMillan Company, 1917), p. 29.
4. Definitions found in the *Universal Dictionary of the English Language,* New York, 1898, p. 4849.
5. In 1708, the historian Oldmixon wrote, "Tobacco is their (Marylanders) Meat, Drink, Clothing, and Money..."Scharf, *History of Maryland,* p. 276.
6. Fletcher, *The Strawberry in North America,* p. 17.
7. *The Sun,* Baltimore, May 17,1918 states that 3,500 to 4,000 families are involved in picking produce from East Baltimore. If we conservatively estimate an average of three persons per family (many families had four or more members involved) the estimate of 10,000 pickers is easily surpassed, even if some pickers went to other counties. Most went to Anne Arundel County during this time period.
8. *The Sun,* June 13, 1910. "Garden of The Cities, Anne Arundel. Where The Berries Come From."
9. *Report of the Maryland Vegetable Growers' Association* by H. H. Zimmerley (U.S.D.A.) published in Report of the Maryland Agricultural Society – The Maryland Farm Bureau Federation, vol xvi, (1931). P. 367.
10. Chapelle, Baker, et al, *Maryland: A History of its People* (Baltimore: The Johns Hopkins University Press, 1986), p. 231.
11. Hamill, W.S., *The Agricultural Industry of Maryland* (Baltimore: Maryland Development Bureau of the Baltimore Association of Commerce, 1934), p. 310 (Table No. 143).

The Farms and the Farmers

CHAPTER TWO: THE FARMS AND THE FARMERS

The roads are blocked,
The pump is froze,
The cider's gone,
And the goose et up,
The chores ain't done.
We're short on hay -
But it's a good life,
Twouldn't trade it
Anyway...

Old Rustic Rhyme[1]

In Anne Arundel County, farming was the primary means of livelihood for almost 300 years. Consequentially, farming methods dynamically reflect the evolution of the modern farm. In colonial times implements were operated by hand and powered by animal or man. As the Industrial Revolution progressed, farm implements were powered by steam, and eventually the gasoline engine was used to drive tractors and large farm implements. This revolution also affected the home life of the farmers with such conveniences as indoor plumbing, electricity, and central heating. The life of the farmer was always a hard life, but substantial changes occurred with the innovations of the Industrial Revolution, which improved the status of the farmer and made farms more productive.

Fairview. Home of truck farmer "Dol" Shipley. Typical I-house, with back wing, constructed around 1830. (Courtesy of Isabel Shipley Cunningham).

Throughout the truck farming period in Anne Arundel County, farms were basically self-sufficient. The farmers produced their own food, as a necessary addition to their truck farming. Most farms had their own garden area for vegetables and fruits, much of which was canned or preserved for future use. This garden area was usually maintained by the women of the household. Most farms had a few cows for dairy products and beef, chickens for eggs and meat, and turkeys and pigs for meat. Each family member had specific chores which contributed to the livelihood of the family. Keeping food on their table was the number one priority.

Road leading to the Cook farm house. Note the typical sandy soil of northern Anne Arundel County. (Courtesy of James B. Calvert).

Most farms had a few acres set aside in forest to provide firewood for fuel as a source of heat and for cooking. Often this forested area was in the middle of his fields which also provided the farmer and his hands a refuge from the summer sun while working. Frequently, these few wooded acres also housed a family cemetery. The farmer did most of the repairs on his home, buildings and equipment, which took much of his time during the winter months. Most farmers even did their own blacksmithing.

Rarely was a doctor available, and farmers had their own medical treatments with tonics, herbs, and other home remedies. Midwives often took the place of doctors.

Horses and mules were used for plowing fields with a single plow. Tractors, automobiles and trucks were not in use until the 1920s. Produce was normally transported to the markets by horse-drawn wagons, sailboats, steamboats or railroad.

Night Soil Tank Wagon. (Photo courtesy of Vegetable Gardening *by Ralph L. Watts, page 51).*

It was the era of buggies and wagons, open wells for water, washing the laundry by hand, and outhouses. No radio, TV or PC's...no telephones, no electricity. The country farms were the last to receive telephones and electricity which started to become a reality for farms around the turn of the century.[2]

The soft, sandy soil of Anne Arundel County that was ideal for the growing of truck produce lacked organic materials. The farmer had to spend much of his winter months preparing the soil for the next growing season. Horses were the common mode of transportation in those days and manure control was a real city problem. But this problem was the answer the farmer was looking for. The "street dirt" from the cities was transported to the farms in boats or in railroad cars and placed on a side rail where the farmers would load it into their wagons and haul the manure to their fields. Farmers would spread as much as thirty tons of manure on one acre. As the use of automobiles replaced horses in the cities, the supply of street dirt diminished and the farmer faced a serious problem. Guano was shipped to Baltimore from the natural deposits of the Lobos Islands in the Pacific. Guano was much more expensive than the "street dirt" and the soil did not respond to guano as well as it had to manure. A few farms used "night soil" (human excrements) transported to the farms on barges from Baltimore and placed into large reservoirs where special tank wagons hauled the material to the farms. Because of the unpleasant character of night soil, it was not widely used in place of other fertilizers, and farmers did not advertise its use. Farmers were charged about $1.50 a ton for night soil.[3] Jeff Cook, at 15, was a rowboss for George A. Jenkins in 1941. He claimed the lack of cheap organic fertilizers, especially "street dirt," was a major contributor to the demise of the Anne Arundel truck farm.

Farmers also needed to add lime to the soil to reduce its acid content. To get lime, oyster shells were obtained from oyster packers or button factories, burned with charcoal and added to the fields. This accounts for the large number of oyster shells still found in many Anne Arundel fields.[4]

An article from the June 14, 1910, edition of ***The Baltimore Sun*** describes in detail the complexities of operating a model Anne Arundel truck farm:

> **$18,000 FROM HIS LAND...Yearly Income Of One Anne Arundel Farmer Is This. $2,000 FROM BERRIES IN ONE DAY "Intensive Farming On An Extensive Scale," Says Owner, Who Here Tells How He Does It.**
>
> *Annapolis, Md. June 13.- As a glorious producer old Anne Arundel county stands far up toward the top of the agricultural heap. In some respects nothing on earth can touch her. Think of what 10,000,000 quarts of strawberries in one season means. That is only one item. No county in the United States produces as much of this fruit as Anne Arundel...*
>
> *What does the reader think of $2,000 worth of strawberries picked in a single day? What will he think of two and in many cases three paying crops of truck raised on the same ground in one season? What will he think of $11,000 in a season from the crop of strawberries alone? "Fairy stories," he will exclaim. And yet they are not fairy stories, but plain, straightforward business records, for the farm responsible for this success keeps books and knows to a cent all the outgo and income from year to year.*
>
> *Let us take one model plantation out of the 4,500 farms that are in Anne Arundel. This shall be Patapsco Plains, generally known as the Hancock farm. The men who own and run it are Messrs. Frank L. Hancock, Lemuel B. Hancock and Harvey L. Hancock.*

Model Anne Arundel Farm

Patapsco Plains, as its name would indicate, is on the Patapsco River. It is near the Solley post office, and its comfortable old mansion can be seen from the river on a high bluff midway between Swan Creek and Cox's Creek. Swan Creek is the first inlet south of Hawkins Point and Cox's is the last before the mouth of Stony Creek is reached. Patapsco Plains has nearly a mile of water front on the river proper.

The Hancock farm consists of 325 acres, of which nearly half is woodland, while 175 acres are tilled on thoroughly scientific principles. It was from these acres that $2,000 worth of strawberries have been picked for market in a single day. This occurred in 1902 and is memorable to the Messrs. Hancock as a banner day, of course, though they have had other days nearly as big when the market was right.

Of this particular transaction it must be explained that the berries did not all go in one shipment. The morning picking is sent from the farm by boat to reach the shipping point in Baltimore at 12:30 o'clock. This lot on the banner day brought $1,069.25. The afternoon's "pick" is sent at night so as to reach Baltimore at 11 o'clock at night to be ready for the early market next morning, and it was this second lot that sold for a little over $930.

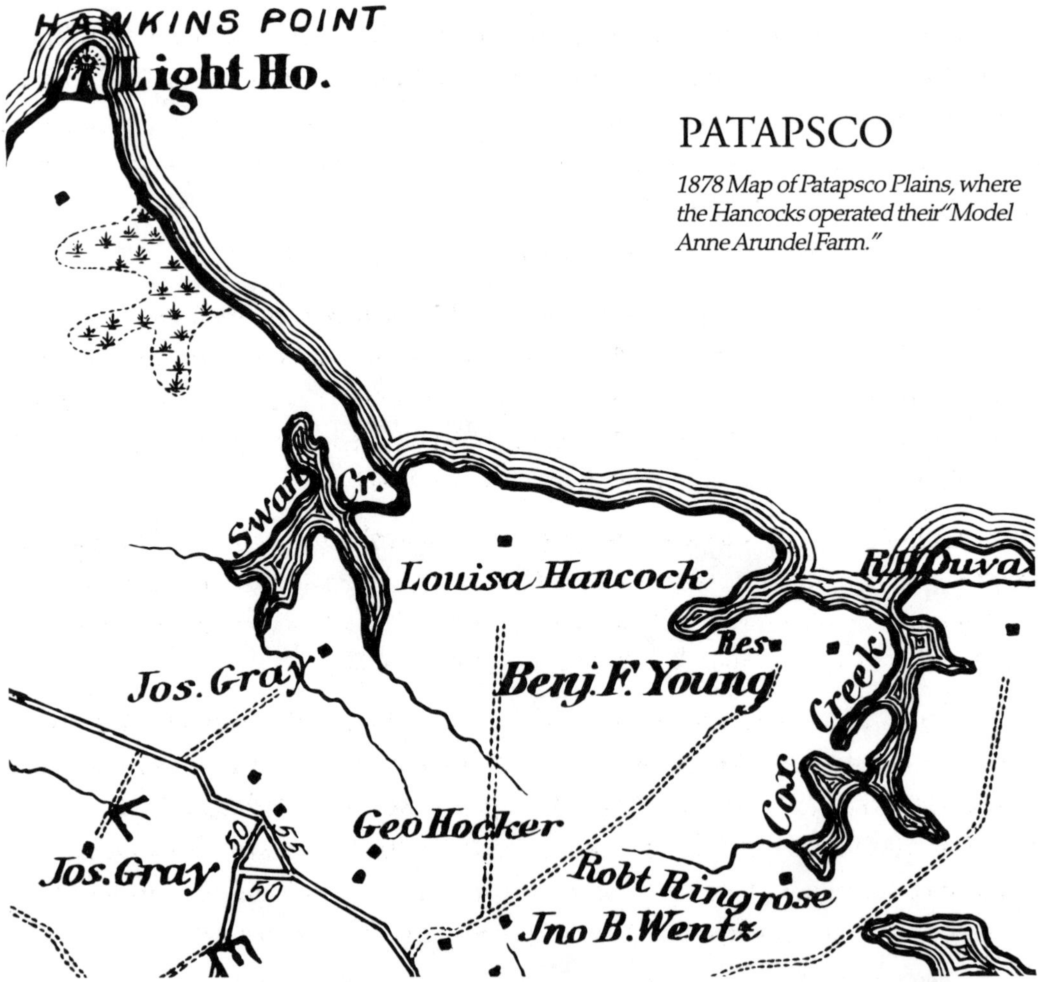

PATAPSCO

1878 Map of Patapsco Plains, where the Hancocks operated their "Model Anne Arundel Farm."

Dead Loads Of Fruit

On this scientific truck farm there are 50 acres of strawberries, and 25 acres in raspberries, blackberries and dewberries. There are 5,000 peach trees, 1,000 apple trees and 100 plum trees.

In addition, an immense crop of tomatoes, potatoes, beans, peas, onions, carrots, melons, cantaloupes and other truck is produced. The whole farm displays a cleanliness and neatness that betokens ample help and minute supervision. No dollar is spared that will add to the productiveness of the land itself, and this goes without reference to whether the Hancock brothers consider a season a good one or a bad one.

In the words of Mr. Frank L. Hancock,

> *"We are practically doing intensive farming on an extensive scale. We find that it pays to keep a comparatively large force of men at work all the winter. It is then that we do everything that is necessary for an early start in the spring. There are many things in the correct conduct of a farm that belong to winter anyhow, but there is mending and repairing of all sorts besides, and it is the sole duty of the man to care for the stock."*

Every inch of crop land on the Hancock farm is worked, even the orchards.

In the case of young orchards the trees are set from 27 to 30 feet apart, thus admitting plenty of sunlight and producing fruit of superior flavor and color. Until these trees get large the orchard is worked in truck crops, consisting of potatoes, tomatoes, beans or peas. None of the plant food that should feed the trees goes to the support of weeds, and besides this the manure put on for these truck crops adds that much more to the growth of the trees.

Cover Crops Always

When the fruit trees grow too large for the easy working of truck the ground is sown in crimson clover, which when in bloom, is plowed under and followed by cowpeas sown by machine in rows 2½ feet apart. The seed is well covered, and all of it comes up. Even in the young orchards after the truck crops come off, the ground is sowed to a cover crop, usually clover, which eventually enriches the soil, since it is the policy here always to have cover crops.

The Hancock orchards easily show the results of winter work. The trees have been regularly sprayed and all loose bark removed to get rid of the harbors of pestiferous insects. The trees are all carefully pruned so as to allow sunlight and air to penetrate. The trunks are smooth, the foliage a strong, healthy green and the branches are loaded with young fruit.

In all the standing fruits like strawberries, raspberries and blackberries, there are selections that produce successive crops throughout the seasons for those fruits. For instance, the list of strawberries runs like this: Climax, Maryland, Nickohmer, Chesapeake, Heritage, Gandy and Stevens late. A peculiarity of the present season is that the earliest berries have to some extent forgotten their function, and are still producing as if they were both early and late varieties. Of the larger varieties this farm has produced many berries six and eight inches around.

Several Crops: Same Ground

The way they raise several crops on the same ground in Anne Arundel is seen at Patapsco Plains as well as on a number of adjoining farms. It is done by the interplanting of the rows. Watermelons will be planted in rows 12 feet wide. Between these will be planted three rows of early peas, all two feet apart. Cantaloupes are set in rows seven feet apart and between these, two rows of peas or beans. The same is done with tomatoes. When the peas are off, other later crops can take their places.

Where peas are planted solidly in fields, as soon as they are out of the way the vines may be used either as hay or they may be turned under as manure and the land then put in tomatoes, field or sugar corn, late potatoes, turnips, rutabagas or squashes.

There are several variations of the three-crop system. For example, rows are made for cantaloupes seven feet apart. These rows are marked across every two feet and where the lines cross, a manure hill is dropped for the cantaloupes. On each side of this cantaloupe hill a raspberry is planted. Thus the order along the cantaloupe rows are marked across every two feet: berry, cantaloupe, two raspberries, cantaloupe and so on.

In other words, along the row laid out for cantaloupes, raspberries go in on the near side and far side of each cantaloupe hill. These rows are seven feet apart. That allows two rows of peas between them that year also. The manure has helped the raspberries wonderfully and the second year the spot is no longer a cantaloupe patch, but a raspberry garden, with still enough space between the rows for a single row of peas or other crop.

Crowds Of Pickers

Truck crops require a great amount of temporary labor in Anne Arundel county. A familiar sight in Baltimore is the wagonloads of Bohemian families who go to Anne Arundel with baggage, boxes, cooking utensils, baby carriages, washtubs and bedding. They go to pick the pea and strawberry fields.

As many as 150 persons are employed in the busiest season on Patapsco Plains, and nearly all of these are men. After years of re-employment these men drift back naturally to the Messrs. Hancock, and there is no dearth of labor. The men are mostly those who at one period are oystermen, at another seine fishermen, at another laborers, and so on. They have their regular rounds and after one job is finished go to the next. Most of them save more money while picking peas and berries than at any other time.

At Hancock's they get no money during the season. They are paid for all work in metal checks, and these are redeemable only on one specified date at the end of the picking season. They do not have to move off the farm for any single item, for on the place is a well-equipped store for their convenience alone. No more than regular country store prices are charged, and the only money passed is the farm checks of brass or aluminum.

Commissary Well Stocked

All sorts of groceries, canned goods, coffee, cheese, canned milk, head cheese, puddings, hams, beef, bacon, simple medicines, clothing, shoes, gum shoes, muslins, cotton, needles, tobacco (20 different kinds), and, in fact, every commodity, except rum, that the men may want can be had at the commissary store. Bread, pies and cakes come fresh from the baker three times a week and the butcher comes three times with fresh meat. In case of sickness or injury among the men a telephone call to Curtis Bay brings Dr. H. B. Horton.

At one time, employers paid the immigrant workers weekly, but the hands used to go to Baltimore for goods, and so many came back with the goods in their heads instead of their hands, that the plan was dropped. If a man comes on the place drunk he gets no job, but many a one who may have been living too high comes a month before his time, knowing he can get a clean bunk, tight roof, and credit for a month. He knows where there's a safe harbor for the fellow who is "dead broke" and he's on the job when it's ready.

In lieu of such a store as exists at Patapsco Plains the Bohemian families usually send a representative to the city to get their supplies, which he transports back to the farm in a wagon.

Neat Market Packages

One object lesson at this model farm is the careful method of gathering and sending fruit to market. Instead of a picker taking an armful of quart boxes to the field to be laid down in the sand until used, and instead of his bringing in the armful of boxes one piled upon the other, causing injurious mashing of the fruit, a better, if a slower, method prevails.

Each man has a light wooden tray with a handle. This holds 10 quart boxes—no more. As each box is filled it is set back on the tray, and the ten are promptly carried to the packer, who stands under a shed. Thus the fruit is packed at once in the crates under shade and no box of berries is in the sunlight of the field longer than it takes to fill nine other boxes. Ordinarily, packing a box is sometimes upset and gathered up again, and the customer finds them gritty. On this farm only bright new packages are used. The crates never come back, and these are worth nearly 30 cents apiece.

This farm, in short, is a model and a fine example of what Anne Arundel can do and has done in making men rich and happy through the generosity of its soil. The Hancock farm is not lonesome in its success, nor is it the only model of its neighborhood, for an immense amount of wealth is represented in the properties of others in this neighborhood.

Business Farmers All

All the large farmers in northern Anne Arundel are applying business methods to agriculture and are doing a business of from $6,000 to $22,000 a year. The Hancock farm makes sales of $18,000 a year. Mr. T. E. Cromwell, on the Annapolis Short Line, near Marley, spent $4,000 for manure and fertilizer this season. Among others who spend annually from $2,000 to $5,000 a year for fertilizers and manure are Hancock Bros., Archer C. Whittemore, Charles B. Pumphrey,

Charles L. Solley, Frank S. Revell, Robert Pumphrey, Thomas. F. Bottomley, C. H. Anderson, Albert Hammond, James S. Benson, James S. Smith, Fulton Kelly, Joseph M. Bryan, Thomas M. Beasley, John Downs and at least 50 others.

In the Third District Mr. Elias Stallings, on the old quarantine farm, put more than $1,100 worth of lime on his land. He is the biggest grower of peaches in the county and in 40 years he has never had a failure in his peach crop.

"These men," said Mr Hancock, "are solid citizens and are among the people who are making Maryland.

Wealth In Transportation

"Their worth runs up into large figures. In the case of Mr. Anderson his farm alone cost $35,000. Nearly every farm has gasoline boats carrying from 40 to 75 tons, and great value in truck is carried in wagons by way of Light Street.

"Did you ever stop to think what these wagons mean in the way of capital? I'll tell you: Just about $800 a team, including the pair of mules, the wagon and double harness. We have men in this section with as many as 10 such teams, which means $8,000, and I know one with 14.

"We have a great advantage by reason of our short haul to market, whether by wagon or private boat, and you must not forget the Rock Creek Steamboat Company, which has two steamers making two trips a day each on some days and three trips on others from Rock and Stony Creeks and Fort Armistead.

Room And Riches For More

"Still we have yet a good deal of waste land and want good men to occupy it. There is no better market in America than that of Baltimore for our products. It is the greatest packing place and has no superior for a shipping market. We have on Marsh Market Space a dozen or 20 men who now load 12 to 20 carloads, when each one of them could load at least 200 cars apiece, because they know the buyers and the advantageous out-of-town destinations where they can place goods.

"Of my own goods, none stop in Baltimore. They are all bought by shippers and go to Chicago, Boston, Pittsburg, Scranton, Buffalo and Rochester. A firm in Buffalo sends a buyer down each year who pays 7 and 8 cents a quart for strawberries for a large fruit juice factory there. After paying this price, together with the freight, it makes a good profit. I cannot see why capital cannot be put into a fruit syrup factory right here in the greatest strawberry district of North Anne Arundel."

As great as the idea may have been, a fruit syrup factory was never established in Anne Arundel County; however, many canneries were established which provided new markets and encouraged the farmers to grow more truck produce.

At its zenith, it is estimated that there were over 2,500 truck farms in Anne Arundel County. Many of the farmers used little brass tokens, called pickers' checks, to pay their pickers. The checks had the farmer's initials and a number stamped on them, designating the amount of produce the picker picked. In the listings of pickers' checks many of the farmers are identified with locations, and biographical notes, if known. Many pickers' checks bear initials which have not been identified, but because of their source are believed to be from Anne Arundel County.

There were some farmers who started their farming careers in Maryland as immigrant pickers and others who were slaves or tenant workers. Some of those known are listed on the following pages.

Immigrant Farmers

Many Polish, German, or Bohemian immigrants eventually realized their dreams by purchasing their own farms. From a Department of Labor study taken in 1923 (see section under Pickers) about one-half of the white farmers surveyed were foreign born. Thus, by the 1920s, a substantial number of immigrants were farming in Anne Arundel County. Some of these were:

Joseph Frank Andrzejewski purchased a farm in 1919 on Dicus Mill Road in Millersville. He had his own pickers and used picker checks until 1956.

Maximilian Blob (1883 - 1969) purchased a farm near Jessup on what is now Max Blob Road. After the demise of the truck farms, the family established Blob's Park which still operates as a center for good German music, dancing and food.

John Blackowicz, died 1960, purchased a farm on Dicus Mill Road. He changed his name to Smith and then converted back to Blackowicz, but retained the nickname of "Smitty."

John G. Friedhofer (1871 - 1931) farmed on land which became a part of Ft. Meade.

Henry Emil Gertz farmed on Chesterfield Road near St. Stephens Church.

Joseph Stanislous Kostkowski (1872 - 1940) emigrated to the U.S. when he was 16. He worked at the fertilizer plant at Curtis Bay before purchasing a farm on Dicus Mill Road.

Benjamin A. Klug, a Pasadena farmer, purchased 30 acres on Mountain Road in 1902. Mr. Klug's farmhouse was later used as a store, post office, and the headquarters of The Rock Creek and Marley Telephone Company

Augustus Carl Schmidt began working on a rented farm near Pinehurst in 1887. He later purchased his own farm on the western half of Riviera Beach.

Louis Schramm, Sr., born in Germany, came to America and became a manager at the Maryland Lime and Cement Company. When he was told by his doctor in 1910, that his environment was unhealthy, he purchased the 236-acre farm of Richard B. Phelps on Mountain Road. Although he had never been a farmer, he learned fast. Mr. Schramm, his wife and five children were soon growing strawberries, peas, beans, peaches, pears, melons and numerous other produce. Their farm became one of the most productive truck farms on the Pasadena peninsula. He later developed the "Schramm" cantaloupe.[5]

Frank Xavier Wimmer, born in Germany and came to Maryland in 1889. In addition to farming, he served as a county commissioner.

African-American Farmers

A substantial number of farmers were African-American. Their role in truck farming has been greatly underestimated or ignored. Some received a piece of land from their former masters when they were freed in 1864, while others saved enough money to purchase their own farms. A large number of the African-Americans involved in truck farming were laborers, working directly for farmers on a year-round basis. Larkin R. ("Dol") Shipley, who farmed "Stamp Act Repealed" in Harmans, had five tenant houses in which African-American families lived and supplied the general farm labor on the large farm.[6] Some African-Americans worked on the farms as tenant farmers, working sections of larger farms, and living in tenant houses furnished by the landowner.

Right: "Dol" Shipley's service truck "Big Bertha" loaded with tomatoes. African-American child on tailgate. (Photo courtesy of Isabel Shipley Cunningham).
Below: African-American family working on "Dol" Shipley's farm. Tenant house on far left. (Photo courtesy of Jack Kelbaugh. Both photos were taken by Claudine Shipley around 1925).

Rachel Hall Brown, born 1912, remembers growing up on her father's farm along the old Annapolis shortline railroad in northern Anne Arundel County. Her father was Rodolphus Hall, called "Dolly", who was born in 1885 and died in the 1940s. Many black farms were located from Marley and Elvaton stations eastward to Solley, Freetown, Spencer's Road and north to Marley Creek to old Spencer's Wharf, where boats ferried the inhabitants to the Baltimore markets. In her mind, this "was truly the garden spot of Anne Arundel County."

Black farms were generally much smaller than those of their white neighbors and Rachel refers to them as "garden farmers" which is technically correct for most Anne Arundel truck farms. Many of their farms consisted of five acres or less. The black farmers often worked on one of the larger white farms in addition to growing their own produce. Many were former slaves and were adjusting to free life.

Rachel Brown recalls the wagons and later trucks carrying their produce into the Baltimore City markets, as well as directly into both white and black residential sections, where fresh garden produce was always welcome.

She remembers many of the black farmers. There were the Spencers, Gaithers, Barleys, Smiths, Franklins, Brooks, Queens, Howards, Hensons, Johnsons, Moses, Williams, Cagers, Turners, Hess, Edwards, Pearmans, Davis, Coates and Rachel's father, Rodolphus Hall. The families often intermarried and she counts her "cousins by the dozens." This was also quite common for the white farmers. Both the Brooks and Turners used pickers' checks.

Rachel recalls many of the black farmers for the produce they grew: Uncle Thomas (Gaither) for his Long John cantaloupes; Uncle Jim (Gaither) for his shiny eggplants; the Turners for their strawberries, which were picked by pickers from Baltimore; Uncle Lige Henson for his green beans; Charles Johnson for his sweet corn and Elizah Howard for his orchard of cherries, plums, and peaches. His wife, Sarah, worked in the garden with him.

The mothers canned vegetables and fruits and made delicious jams, spreads and preserves. The canned food sustained the families during the winter months. Rachel's grandfather, Joseph Burley, was known for his beautiful garden filled with watermelons. He planted special rows for his grandchildren and designated a special fruit tree for each grandchild. He also grew turnips and asparagus which the children called "grass" and refused to eat.

These hard-working garden farmers were the first black entrepreneurs in the county. They managed to feed their families during the depression and were instrumental in the establishment of black churches and schools.

Rachel and many of her cousins still get together on Labor Day at Magothy Camp Meeting to reminisce over their treasured days of garden farming in northern Anne Arundel County.[7]

Many children worked during the picking season to earn extra money for their families. In 1923, The Department of Labor published a study on Child Labor on Maryland Truck Farms. Most of the research was conducted on Anne Arundel farms. The study reported that, of the 540 children who worked on the farms and lived in the area, 322 were African-American.

African-American farmer with sweet potatoes.
(Photo courtesy of Ann Arrundell County Historical Society).

We learn much about the African-American farms of the 1920s from this study. Their farms, averaging about 20 acres, were considerably smaller than those of white farmers, averaging over 50 acres. The African-American farms were for the most part in regions where waste lands, swamps, and woodlands were more prevalent than in defined farming districts. Most of their homes were unpainted cabins of three or four rooms with outdoor privies and a water supply that came from shallow wells, springs and brooks. About two-thirds of the African-American farmers were literate.[8]

Some of the known African-American farmers who operated their own farms and used their own pickers' checks were:

> **John Henry Robinson**, born a slave in 1844, served as a house boy for Basil Smith Benson for whom his mother was a cook and wet nurse. John Henry was freed in November 1864 under the new Maryland constitution. He became a coachman and footman at Carroll Manor and married Della Queen, a former slave in the Linthicum home of Sweetzer Linthicum. John Henry borrowed money from G. Milton Benson, the eldest son of Basil Smith Benson, and purchased land on the Baltimore County side of the Patapsco River. He built a house and barn and raised corn, vegetables and fruit. In addition to farming, John Henry stripped willows growing along the banks of the Patapsco and made wicker furniture. Their home was beloved by all the residents of nearby Halethorpe, and the local kids were always treated by Aunt Della and Uncle John to goodies from the farm or Aunt Della's homemade pie or bread. John Henry Robinson died in 1931.[9]

John Henry Robinson, African-American truck farmer.
(Photo courtesy of Ann Arrundell County Historical Society).

Ulysses W. Brooks farmed on Furnace Branch Road near Glen Burnie.

Henry Brooks also farmed on Furnace Branch Road. He was the father of Urias Brooks (1877 - 1964).

William Ware farmed near Ordnance Road.

Nick Simms farmed on Race Road in Hanover.

Summerfield Jackson, 1863 - 1945, farmed in Hanover.

William C. Dotson, 1841 - 1927, was born a slave. He farmed on Furnace Branch Road in Glen Burnie. After becoming a free man, he changed his name from Dodson to Dotson to overcome his self-image as a slave. His land was later farmed by his two sons, Danial C. Dotson and Richard James Dotson. The sons punched "RR" on the reverse of their father's checks.

George Williams, 1822 - 1890, farmed near the intersection of Snow Hill Lane and Cedar Hill Lane in Brooklyn. His two sons, Rudolphus Williams (1856 - 1930?) and Richard T. Williams (1858 - ?) were also truck farmers and used their own checks.

John Henry Carroll, 1839 - 1907, was born a free man, but his mother was a slave. He farmed in the 8000 block of New Cut Road.

John Barnett has been farming in Tracy's Landing for over 80 years. He gave up farming tobacco in the 1960s and has been growing "a little bit of everything" since. In 1999, he was still selling his produce at the farmers' market on Riva Road.[10]

There were undoubtedly many more African-American farmers, but there is little documentation. More information will certainly come to light in the future to add to the significance of the African-American's role in truck farming.

John Barnett at the Farmer's Market on Riva Road. (Photo by author).

Trucking in South County

While a majority of the truck farmers were north of the South River, many farmers in the southern part of Anne Arundel County dedicated part of their land to truck produce. As early as 1871, strawberries were grown in the Eighth District, which was located at the extreme southern tip of Anne Arundel County.[11]

Many large peach orchards were planted in the Friendship area and the Harris Orchard in Lothian, with approximately 2,000 peach, apple and plum trees, still grows strawberries, blackberries and raspberries. In the 1920s, on one day, over 30,000 baskets of tomatoes were shipped from Shady Side to the Baltimore markets. During the summer, oystermen used their oyster boats to haul watermelons and other vegetables to Baltimore and Annapolis from West River. South County farmers concentrated on their tobacco harvest, but many contributed to the produce markets in Annapolis and Baltimore by adding tomatoes to their tobacco crops. Many seafood canneries operated in the West River area and a cannery was established in Friendship in 1888 for the primary purpose of canning fruits and vegetables. [12]

In 1922, the first Anne Arundel strawberries to reach the Annapolis markets came from farms in the West River area, and "were large, luscious and sweet."[13] William Witt, an immigrant from Germany, farmed in Mayo. He grew watermelons, tomatoes and other farm products and took them by boat to Baltimore. Blanch Brashears grew tobacco and tomatoes. Captains Sam and Will Carr at one time produced large quantities of delicious watermelons on their farm. When their melons were ripe, they would load them into a boat and sail to Shady Side where they were sold to Captain Bob and Matilda. The areas of Beverly Beach and Coverlea were once farmed for wheat and tomatoes. Captain Bowie Cove grew tomatoes and asparagus on his farm at Turkey Point.[14]

Steamer Emma Giles *loading at Taylorsville (Riva), Beard's Creek, South River. Farmers brought their wagons to the docks where their produce was shipped to Baltimore. July 1917. (Photo courtesy of Mrs. William Luebbers).*

Steamer Emma Giles *(Postcard in author's collection).*

In 1891, the steamboat *Emma Giles* was scheduled to make the Annapolis, South, Rhode and West River run. Steamboat days were Mondays, Wednesdays and Fridays and, during the summer, Saturdays and Sundays. "The wharf was piled with hogsheads of tobacco, or depending on the time of year, bags of wheat, baskets of peaches and tomatoes, seafood and fish packed in ice to be loaded aboard the *Emma*, with a backup of wagons and carts." *Emma Giles'* last run was in 1932.[15]

Quite a few of the farmers from the southern part of the county gave up tobacco in the 1950s and 60s and turned to truck produce and other grains. Several of these farmers, such as the Zangs from Galesville and John Barnett from Tracy's Landing, are found in today's farmers' markets selling their vegetables and fruits.

Camp Meade

In June 1917, the Secretary of War announced the selection of Admiral, Maryland as a site for one of 16 army camps to be built to train soldiers for World War I. Nineteen thousand acres of mostly farmland and forests were needed to build the training camp. Farmers and landowners were notified by mail to attend a meeting at Admiral in a W. B. & A. railroad car, where they were told their land would be purchased by the government and it would be unpatriotic to protest. Over 600 transactions were required for the government to obtain the land at an average price of $81.15 per acre, which included homes and outbuildings. The first tract of land of 15,000 acres was to be vacated by October 1, 1917. The farmers were not allowed to harvest any crops and many tomatoes, melons and vegetables were left to rot in the fields.[16]

Leaving their farms was very difficult for the many farmers. Penrhyn S. Watts was born on his father's farm in Admiral. At the age of 16, he dropped out of school and took over the farm when his father, James P. Watts, died in 1906. Penrhyn was forced to sell his farm to the government in 1917. After trying other work, he went back to farming in 1922 when he purchased a farm in Severn on Old Telegraph Road. He was noted for his large sweet strawberries, which would fill a box with six or seven berries.[17]

Many farmers, displaced by Camp Meade, purchased farms in other areas of the county, but many never farmed again.

By the end of the 1920s, truck farming was on a decline in Anne Arundel County. Refrigerated trucks and railroad cars were bringing produce to the northern markets before local produce was ripe. Land prices were escalating rapidly and with the great depression, many farmers sold their land to developers.

World War II revived many farms as food was in big demand to support the war. However, after the war, construction projects absorbed more farmland and truck farming almost completely disappeared from the county. Many farmers turned to tobacco for their cash crop, which accounts for the many large tobacco barns, still standing throughout the northern part of the county. The construction of Friendship Airport, new highways, housing and commercial developments displaced many farmers and absorbed thousands of acres of land.

Friendship Airport

In 1948, the City of Baltimore purchased 3,200 acres of land in the Friendship area in northern Anne Arundel County to build a city airport. The airport opened in 1950 as Friendship International Airport which later became known as Baltimore/Washington International Airport (BWI). Again many farmers were forced to leave the land their families had farmed for generations. The Benson-Hammond House is the only old farm building that survived the construction of the airport.

Construction and Taxes

Numerous farms have been devoured by construction projects. The Baltimore beltway and highway construction has absorbed thousands of acres of land. Many farms were consumed or divided by highway construction and numerous farmers have left farming. The price of land had risen so high that they could not afford a new farm.

Adapting to changing times, in 1946, the Schramms on Mountain Road began raising turkeys. Two years later they opened a roadside stand selling vegetables, fruits and flowers. Although reluctant to sell their farm, large estate taxes forced them to sell most of their farm in 1991. They have retained about 35 acres so they can continue raising a limited amount of truck produce.[18]

Remnants of Truck Farming

In 1992, there were 477 farms with over 43,000 acres still operating in Anne Arundel County. Corn, soybeans, hay and wheat are the major crops along with some oats, barley and tobacco.[19] Very few truck farms are left. In 1988, six Anne Arundel County farms offered "Pick Your Own" services but in 1999 only two farms were listed.[20] One of the largest truck farms still in existence is operated by Jimmy Schillinger on New Cut Road adjacent to I-97. It is called "Papa John's" after Jimmy's father, John Schillinger who, in 1937, took over the farm that was started by his wife's grandfather, James F. Wade, in the 1870s. The Schillinger's farm of approximately 375 acres grows a wide variety of fruits, flowers and vegetables which include tomatoes, peppers, eggplants, cantaloupes, watermelons, sweet potatoes, cabbage and pumpkins. They have also reintroduced strawberries on a limited basis. They market most of their produce to local supermarkets and operate a roadside produce stand made from an old tobacco barn. They also sell their produce in the farmers' markets. John Schillinger died in 1995, but the family continued operating the farm. With sophisticated irrigation systems, the farm has been very productive. Several generations of Schillingers are actively involved and they have no plans to give up the farm.[21]

Ruth and "Papa" John Schillinger. (Photo courtesy of Jimmy Schillinger).

CHAPTER TWO NOTES

1. *Report of the Maryland Agricultural Society, the Maryland Farm Bureau Federation*, vol xvl, (1931). Address by Dr. R. A. Pearson, President of the University of Maryland. Adaptation of an old rustic rhyme.
2. Shipley, E. Roderick, *Truck Farms in North County*, published in *Anne Arundel County, Maryland: A Bicentennial History 1649-1977*, Annapolis; Anne Arundel County and Annapolis Bicentennial Committee, 1977, p. 68. Good reference for a more detailed description of life on a truck farm.
3. Watts, Ralph L. *Vegetable Gardening* (NewYork: Orange Judd Company, 1914), pp. 50-52.
4. Shipley, E. Roderick. *Truck Farms in North County*, pp. 71-72.
5. Cunningham, Isabel S., *Between Two Rivers* (Pasadena: Pasadena Business Association, 1999), p. 26-34.
6. Cunningham, Isabel S., *The Recollections of James W. Shipley: Growing Up on the I. L. Shipley Brothers Farm*, Anne Arundel County History Notes, Vol. XXVI, No. 3, p. 13.
7. Information taken from a written statement by Rachel Hall Brown, March, 2000.
8. Channing, Alice. *Child Labor on Maryland Truck Farms*, U.S. Department of Labor, Children's Bureau, Bureau Publication No. 123, Government Printing Office, Washington, D.C., 1923.
9. Dorothy Bensen Keen, "John Henry Robinson," *Anne Arundel County History Notes* 17, no. 2 (Jan., 1986).
10. Personal conversations with John Barnett, September, 1999.
11. *The Maryland Republican*, April 29, 1871, "Eighth District."
12. *The Evening Capital*, April 28, 1888, "A Cannery at Friendship."
13. *The Evening Capital*, May 17, 1922, "First Strawberries of County Retailed at 25 cents per quart."
14. Mullins, Caroline L. Britt, *The History of Mayo, Maryland (Baltimore:* Gateway Press, Inc., 1996), pp. 135-154.
15. Dunham, Dorothy Lee, *Galesville, Maryland: Its History and its People (Severna Park:* The Paper Mill, 1980), p. 30.
16. O'Malley, Catherine L., *Odenton, The Town a Railroad Built* (Odenton: Revere Printing, Inc., 1987), p. 37.
17. Interview with Margaret Rose, daughter of Penrhyn S. Watts, Jan. 4, 2000.
18. Interview with Emma Schramm, Mar. 5, 2000.
19. Bureau of the Census, U.S. Department of Commerce. Data was issued in a handout (1998) by The Maryland Department of Agriculture, Marketing Resource & Development Group, Annapolis, MD, 21401.
20. From Maryland Direct Farm Market & Pick Your Own Directories of 1988 & 1999. Compiled and edited by: Maryland Direct Farm Market Association. Available from the Maryland Department of Agriculture Office of Marketing and Agricultural Development, Annapolis, MD 21401. Listed in 1988 were: Boyer Farms, 1500 Severn Road; Gambrills Orchard Farm Market, 1400 Waterbury Road; Holt's Produce Farm, 4808 Mountain Road; Mt. Airy U-Pick Farm, 832 Mt. Airy Road; Pumphrey's Home Grown Vegetables, 10 Rt. 3 South; Wilton's Organic Plants, 357 Hadlem & Catherine Avenues. The 1999 Directory lists only two farms: Boyer Farms and Pumphrey's Home Grown Vegetables.

    In recent years Strawberries could be picked at: Pettebone Farm, 1576 Bayhead Road; Hearon Buttrill, 1616 Meeting House Road; Louis A. Boehm, Jr. (Country View Farms), 1369 St. Stephens Church Road; Mt. Airy U-Pick Farm, 832 Mt. Airy Road; Leonard Kinder, 2310 Davidsonville Road; Robert Robey, 132 Waugh Chapel Road; Audrey Paddy, 631 West Bay Front Road; Jimmy Schillinger, 8029 New Cut Road.
21. *The Capital*, June 17, 1995, "Papa John Schillinger Dies: Ran Farm 58 Years."

Strawberries and Other Produce

CHAPTER THREE: STRAWBERRIES AND OTHER PRODUCE

"This is Anne Arundel"

This is Anne Arundel - a lady in the spring
Picking ripe strawberries in the patches of her dream.
All along the Short Line those blooms and berries lie -
And May is in the willows with a song of never-die.
This is Anne Arundel - all in her girlish state
Gone out to cut asparagus and swing upon the gate!

There's young fruit in her orchards,
there's music in her eyes,
There's magic where her Severn stream in azure beauty lies.
A mist o'er Gibson's Island, the Magothy a bay,
And peas are ready to be picked - but never mind today!
This is Anne Arundel - an old land turned to new
With roses in her honey'd lanes soft-kist with magic dew.

She holds the mighty Navy upon her lap of stars,
This sister of the peace that mocks the agony of Mars.
And down her roads of beauty so many shadows stroll
That are the ghosts of many a fine, superb Colonial soul:
This is Anne Arundel - a strawberry land of joy,
With cherries in her rosy cheeks to
tempt some wandering boy.

The Bentztown Bard (Folger McKinsey)
The Baltimore Sun, May 15, 1929

Strawberry Culture

The early American colonist found an abundance of wild strawberries in the new world. No matter where they settled, Virginia, Massachusetts, or Maryland, they found strawberries, "four times larger and much more exquisitely flavored" than the wild strawberries of England.[1]

Strawberries have been known to man as long as history is recorded. Cato, the Roman senator (234-149B.C.), was especially fond of the small red berries, soaked in wine or sprinkled with salt and spices. The poet Virgil (70-19BC) called the strawberry, "the fragrant fruit born of the soil." Nor did the fragrant fruit escape the quill of Shakespeare as he quotes King Richard III speaking to Morton, the Bishop of Ely: "When I was last in Holborne, I saw good strawberries in your garden there. I do beseech you to let us have a mess of them..."

Charles V of France purchased 12,000 strawberry plants in 1368 for the Royal gardens in the Louvre. Europeans considered strawberries an expensive delicacy found only in the gardens of royalty.[2]

In the new world, strawberries grew everywhere. As fast as settlers cut down the forests, strawberries appeared in the clearings. Abandoned fields soon abounded with wild strawberries. Very few colonists bothered to cultivate the berries as they were so plentiful there was no need to transplant them. Occasionally an industrious planter would have a few of the superior plants transplanted to his garden. Most fruits were imported from Europe, but the American strawberry was far superior to any variety grown in Europe, and the American berry was introduced to England. There it was crossed with a large strawberry cultivated by Native Americans in Chile. The result was a large-fruited berry that is an ancestor of our strawberries. Shortly before 1800, the improved varieties were being sold by American nurseries, making commercial cultivation near large cities possible.[3]

Long before the American Revolution, wild strawberries were used as an article of barter and were sold in the Baltimore markets. In those early days strawberries, as well as many vegetables and fruits, were confined to the home gardens of gentlemen of means who employed private gardeners. The cultivation of strawberries for market did not begin until after 1800. Nurseries began to develop several new varieties of strawberries that were far superior to the wild berries. The demand for the new larger berries far exceeded the supply and, in response, commercial strawberry culture began.

As early as 1815, perhaps sooner, Anne Arundel County began shipping strawberries to the Baltimore and Philadelphia markets. One hundred years later, Maryland, which laid the foundation of the strawberry industry, maintained preeminence in strawberry acreage.[4]

A Maryland correspondent to the Dec. 26, 1823, issue of *American Farmer* reported a technique of growing native strawberries from the woods that yielded the "enormous quantity of 80 bushels" per acre. He had his results verified by "several gentlemen of high respectability" as the average yield in those days was about 30 bushels per acre.

The *Maryland Gazette* on May 27, 1830, reported that Annapolis strawberries were shipped through the new Chesapeake and Delaware Canal and sold for fifty cents per quart in Philadelphia.

Before 1840, Richard Crisp, a farmer from Kent Island, purchased land in Anne Arundel County and planted a large strawberry bed from which he sold the berries in Baltimore and Philadelphia for an excellent price. These early transactions gave birth to the commercial strawberry industry in the United States.[5]

Anne Arundel County, by 1850, was the most important strawberry district in the South with over 250 acres dedicated to the fruit.[6] The June 9, 1852 edition of ***The Baltimore Sun*** ran the following:

> ***The Strawberry Trade.*** - *Few persons in this city, and much fewer beyond it, have an idea as to the extent of the strawberry trade between our gardeners and the Northern markets. The amount of the delicious fruit sent to Philadelphia, by the Ericsson line of steamers average fifty thousand quarts daily. These are nearly all raised on farms in Anne Arundel county, on some of which forty to fifty acres are devoted to their growth, and 150 to 200 hands employed by each proprietor in picking them.*

In 1854, there were approximately 1400 acres of strawberries under commercial culture in the United States and Canada. About one-third of those acres were found in Anne Arundel County.[7]

By 1857, the county had over 600 acres of strawberries which were producing 20,000 bushels. The "Strawberry King" of Anne Arundel was Rezin Hammond, who planted over 100 acres of strawberries on his farm a few miles south of Brooklyn on Hammonds Lane.[8] In 1859, Thomas Meeham reported in the *Gardener's Monthly* that,"The freight train from Baltimore to Philadelphia on the night of the 7th of June brought upwards of 60,000 boxes" of Anne Arundel strawberries.[9]

For the next hundred years, Anne Arundel County was captive to a strawberry culture which stained the fingers and mouths of every citizen. Strawberries were the first of the spring produce to ripen and marked the true beginning of the vegetable and fruit cycle each year. By the end of May, strawberries were in full bloom and graced the countryside and dining tables with their delicious taste and sweet fragrance. Anne Arundel County was immersed in a strawberry culture which they shared with a large part of the country.

"Strawberry Fever" swept the United States between 1858 and 1870.[10] Even the Civil War did not divert strawberry enthusiasts. With the early capture of Norfolk, strawberries and peas were shipped to Baltimore by steamboat in May, 1862.[11] ***The Baltimore Sun*** on May 26,

Strawberry Vendor on Maryland Avenue, Annapolis, 1893. (Photo by Frances Benjamin Johnson taken for the 1893 Columbian Exposition at Chicago, Special Collections, Nimitz Library, U.S. Naval Academy and Robert Merrick Archive, Maryland State Archives).

1864, in the middle of the war, ran three classified ads for 280 women to cap strawberries. The luxurious berries commonly sold for thirty to forty cents a quart and an acre of strawberries would often realize $1,000 profit. With these high profits, grain and stock farmers, merchants, and speculators rushed into the berry business.

Overproduction by 1870 caused a sharp decline in prices to three or four cents a quart, which was well below production costs. Many farmers lost heavily, especially those who had speculated in the business. The years 1870 to 1885 are known as the "slump" period for strawberry growers; yet most veteran growers in Anne Arundel County survived the great slump and may have benefited from it as the slump drove the speculators out of the business.[12] During the boom years, strawberries were introduced to the Eastern Shore of Maryland and with the construction of the Delaware Railroad in 1865, northern markets were readily accessible. By 1888, despite the slump, over five million quarts of strawberries were shipped north over the Delaware Railroad.[13] However, Anne Arundel County still retained its number one ranking with the largest production of strawberries. By 1910, Maryland had more acreage planted in strawberries than any other state in the nation and Anne Arundel County, with 3,937 acres of strawberries, led all Maryland counties in the production of the delicious berry. [14]

The Strawberry Festival

Strawberry Fever gave rise to the strawberry festival which was annually held during strawberry season to raise funds for good causes. Strawberry festivals were held as early as 1858, probably before, and became an integral part of this culture.[15] Throughout the later part of the 19th century and the early half of the 20th century, strawberry festivals were held as fund raisers by many churches and civic organizations. Ice cream, cake, and sometimes homemade candy were served with the delicious berries for an admission price of 10 to 25 cents. Entertainment and music were often provided. The Annapolis High School classes of 1909 and 1910 held strawberry festivals to help defray commencement expenses. In 1910, unfavorable weather kept the receipts low and the students repeated the festival on the next night.[16] In addition to the church groups, festivals were held by the YMCA, Epworth League, Independent Order of the Rachabites (Eastport Tent No.1099), Fraternal Order of Eagles, Helpers League, Ladies' Aid Society, Garden Farm Improvement Association, and many other organizations.[17]

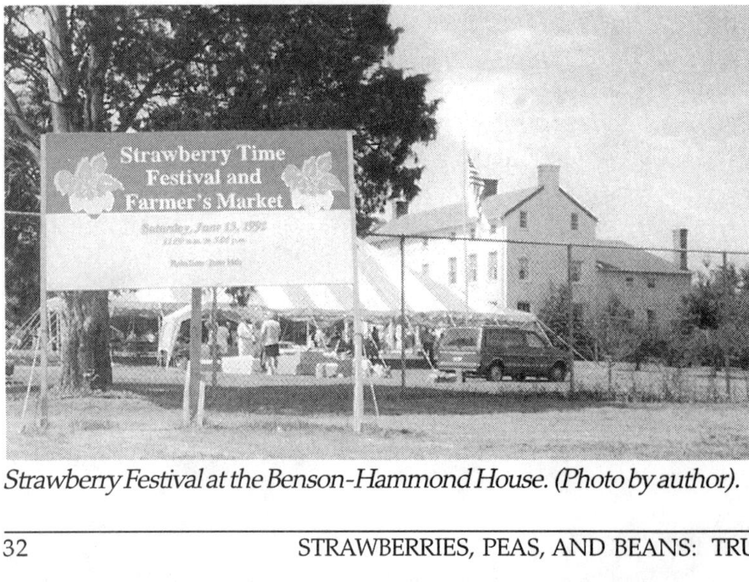

Strawberry Festival at the Benson-Hammond House. (Photo by author).

So many church groups held festivals that, in 1913, the Rev. Dr. D. H. Steffens, a Lutheran minister, insisted that overemphasis upon the social attractions of churches had made churches worldly. "People are making 'sociability' the condition of their membership... they are forgetting that the only work of church really enjoined upon it by its Lord is the evangelization of the world through the preaching of the Gospel."[18] His objections seem not to have deterred the churches from their annual festivals as they continued through both World Wars and the depression years. In 1942, a festival scheduled at Trinity Methodist Church in Annapolis became a casualty of the war as women from the Society of Christian Service were unable to get sugar for the event and the festival had to be cancelled.[19]

The strawberry festival lingers in Anne Arundel County. The Davidsonville United Methodist Church still offers fresh-picked local berries at their festival held every year. Baldwin Memorial United Methodist Church on Generals Highway, the Ann Arrundell County Historical Society, and other organizations in the county still hold strawberry festivals, but to their chagrin, most of the berries are now imported from California!

Strawberry Festivals were often advertised. The earliest found is from ***The Sun*** (1860) others from ***The Evening Capital***:

STRAWBERRY FESTIVAL.—There will be a STRAWBERRY FESTIVAL in the Basement of Exeter Street Church, (instead of Marion Hall, as printed on the Tickets,) on THIS (FRIDAY) EVENING, June 8th. The proceeds to be applied to the support of the Northeastern Home Mission.
TICKETS 15 CENTS. je6-3t*

First advertisement of the Strawberry Festival found in area.
The Baltimore Sun, *June 8, 1860.*

Strawberry Festival.—The ladies of the Presbyterian church, contemplate holding a Strawberry and Ice-cream festival the coming week at the Assembly Rooms. The funds derived from the enterprise, will go toward the improvement on the church. It is to be hoped they will be liberally patronized. Due notice of the time will be given in hand-bill.

Maryland Republican, *June 5, 1875.*

THE SENIOR CLASS
ANNAPOLIS HIGH SCHOOL
WILL HOLD A
Strawberry Festival
At the City Assembly Rooms,
Friday Evening, May 21st, from 7 to 10
And Saturday Afternoon, May 22nd,
From 4 to 6, for benefit of the Children.
Tickets, 25cts.
entitles holder to Ice Cream, Cake and Strawberries; or Admission Free.
Liberal patronage solicited. Proceeds to defray commencement expenses. 18

The Evening Capital, *May 21, 1909.*

Festival and Lawn Party.

A Lawn Party and Strawberry and Ice Cream Festival will be given by the Ladies Mite Society at the First M. E. Church and Parsonage Lawn adjoining, *Thursday Evening May 29th.* The ladies having the matter in charge promises an enjoyable evening to all who attend. The public are cordially invited to be present.

The Evening Capital, *May 27, 1890.*

STRAWBERRY FESTIVAL
St. Luke's Hall, Eastport
WEDNESDAY, JUNE 5th, 7 P. M.
Berries, Ice Cream and Cake
25 cents

The Evening Capital, *June 1, 1940.*

Strawberry Festival

Davidsonville United Methodist Church will hold its annual Strawberry Festival from 11 a.m. to 3 p.m. June 6 at the church, Routes 424 and 214.

The menu will include strawberries, ice cream and shortcake. Additional refreshments and a bake table will be available.

All proceeds will benefit the youth program and go toward helping the youth participate in the Appalachian service project this summer.

The Evening Capital, *May 24, 1998.*

Contemporary newspapers colorfully reflect the Anne Arundel strawberry culture.

From ***The Evening Capital***, May 27,1889:

THE BONES IN THE SHAD

"When the angels made shad,
The devil was mad,
It seemed such a feast of delight;
So to spoil the fine scheme,
He jumped into the stream,
And stuck the bones in out of spite.

When the strawberry, red,
First illumined its bed,
The angels looked down and were glad;
But the devil, 'tis said,
Fairly pounded his head,
For he had used all his bones on the shad."

Anonymous

The shad was one of the most popular fish for eating from the Chesapeake Bay. Both the shad and the strawberry are now close to becoming relics of the past. Efforts are now underway to bring the shad back to the bay, but there is little hope of bringing back the strawberry.

From ***The Evening Capital***, May 19, 1894:

A NOVEL STRAWBERRY BED

Mr. B. F. Baker, superintendent of the national cemetery, has one hundred and forty strawberry plants growing in and around an ordinary flower barrel. The barrel is filled with rich, mellow earth, and on the outside are numerous auger holes, about an inch in diameter. From each of these holes is a flourishing plant, filled with fine luscious berries, while the top of the barrel is also covered with the plant all in a healthy and flourishing condition. From the looks of this novel "strawberry bed," Mr. Baker will realize several quarts of fine berries. The idea is an original one with the superintendent, and besides the novelty of the thing, is quite an ornament to the grounds. Mr. Baker says he expects to have a number of such "strawberry patches" next season.

Growing strawberries from a barrel appears to have been a fad in the late nineteenth century. Up to forty quarts of berries could be grown from one barrel and the strawberries ripened ten to fourteen days earlier than those in the field. Barrel strawberry culture failed more often than it succeeded and was an interesting novelty for the amateur.[20]

From ***The Evening Capital***, May 28, 1890:

Yes, we will take strawberries, green peas, cucumbers, green apples, or anything that is handy for back subscription.

From ***The Evening Capital***, May 28, 1884:

Long Dock, Baltimore, presented a scene of business activity by the first large arrival of boats from the truck farms of this county. They were freighted with large quantities of luscious strawberries which were sold at good prices. The Sun says: a feature of the truck fleet was the handsome appearance the boats made. All have been recently painted in bright colors. Many were decorated in bunting in honor of their first trip with the season's products. On the return trip most of the boats carried down bands of pickers, consisting of men, women and children. The arrangements of last year, by which there were night and day boats, will be carried out this year. This is done to avoid the great inconvenience that would grow out of so many vessels being in the dock at one time, as well as afford a better chance for shippers to get their truck off for local points by early trains, or have it in market in the morning.

From ***The Evening Capital***, May 29, 1889:

STRAWBERRY CROP
Some Facts Connected with Its Growth

The strawberry crop, on account of the frequent rains, may prove less profitable than anticipated by grower and dealer. As far as can be learned from our county growers, from whom Annapolis gets its principal supply, the berry crop early gave promise of abundance. During the period of blossoming the weather was propitious and the outlook for an abundance of the luscious fruit was exceedingly encouraging. But all this was changed by the frequent heavy rainfalls, which to a greater or less degree, blighted the fair indications and caused the hopes of the grower to sink several degrees within him.

Persons unacquainted with the extent of the strawberry business transacted in this city during the season, which will soon open with all its briskness, will be loth to believe that thousands of boxes of this delicious berry are consumed by the people of Annapolis besides what goes to Baltimore, Washington and other cities. It is estimated that not less than 15,000 to 20,000 boxes of strawberries are bought and sold each season in our market. At an average price of ten cents per box, the sum of $1,500 to $2,000 is realized from strawberries alone. The profits in the berry trade, belong principally to the grower. To the average dealer, the losses, in many instances, overbalance the profits. Dealers unacquainted with the perishable nature of succulent fruits will realize sooner or later that their motto must be one that is attired in the cobwebs of antiquity - "Quick sales and small profits." However, the strawberry season is one of profit to a great many people, and the differences that now and then bubble up above the surface of everyday business life, are but the natural result of competition between legitimate trades - But let the strawberry crop come to us at prices within the reach of all; let the milkman continue to serve us with pure and unadulterated *cream, and the grocer dispense pure sugar, then we proud Annapolitans may be happy once more—The manager of the seductive festival may smile with satisfaction, and the young man and maiden go down the path of life hand in hand, happy in their endeavor to kiss from carmine lips the stain of the favorite Anne Arundel strawberry.*

From ***The Maryland Republican & Advertiser***, May 27, 1871:

Our farmers are busy gathering and hauling strawberries...One of the largest growers is Joseph M. Brian, Esq., who expects to pick 80,000 quarts.

From ***The Maryland Republican & Advertiser***, May 20, 1876:

The first strawberries of the season were picked on last Saturday by Mr. Cyrus Joyce, and readily brought forty and fifty cents per quart. The full crop will not be ready for market before the 24th, at which time it is expected a large quantity will be shipped.

The "Cyrus P. Smith" has been engaged to run up the Severn River during the fruit season, and will begin her regular daily trips on Monday, 22nd.

From ***The Maryland Republican & Advertiser***, June 14, 1886:

FINE STRAWBERRIES

One of the members of the firm of W. L. Snyder & Brother, the well known fruit dealers, brought a crate of strawberries raised by Mr. Wm. H. Hall at Curtis' Creek, Anne Arundel County, to **The American** *office yesterday. The berries are known as the "Monarchs." They are of excellent flavor and the size of an ordinary peach. This is the most liberal fruit grower we have ever heard of in Anne Arundel. Two or three boxes is generally the limit to most editors, but* **The American** *being the largest paper in Baltimore accounts for this unequal share of berries.*

From ***The Evening Capital***, June 3, 1889:

John Gumpman, aged 37 years, residing at 606 Port Alley, Baltimore, was brought to the city on a tug last night from Rock Creek, Anne Arundel County, suffering from gunshot wounds in his right leg and arm. Gumpman, it is alleged, trespassed on the strawberry patch of Edwin Hines, and while picking the fruit was shot by Hines, who was in ambush. His wounds are slight. Gumpman's hat was pierced by the shot.

Edwin J. Hines, truck farmer and sharp shooter. (Photo courtesy of Ann Arrundell County Historical Society).

From ***The Evening Capital***, June 3, 1891:

SHIPPING ANNE ARUNDEL STRAWBERRIES

That the Anne Arundel County strawberry is appreciated in other places than the vicinity of its growth, says the Baltimore Sun, is shown by the large shipments of this fruit to points throughout the North and West. On Saturday there was shipped by the Northern Central Railway from Calvert Station, twenty four carloads of berries and nine from President Street Stations. The berries go to Buffalo, Syracuse, Rochester, Utica and as far north as Toronto, Canada. The shipments are also made to Cleveland, Pittsburgh and Chicago. Fruit is packed in refrigeration cars, each car averaging about 150 crates. From four to five tons of ice is placed in each car to keep the berries. These trains leave Baltimore at 7 P.M., and reach their farthest destination not later than 4 A.M. on the second day after shipment. The Northern Central has made a specialty of this business, and by giving fast service has aided materially in building up an outlet for the fruit growers of Anne Arundel. The berries are gathered in the early morning and brought to Baltimore in large quantities during the day. By the care taken in packing them they reach their destination in a wholesome condition. The Eager-Street yards, where the berries are received and loaded on the cars, is a busy place during the arrival of the fruit. The packers and buyers gather there in force and carry on a general trading business in the product. It is expected that these large shipments will continue until Saturday, when the Northern berries will begin to ripen and the demand for Southern fruit falls off. It is said that the berry crop of Anne Arundel is unusually large this year.

From ***The Evening Capital***, June 5, 1893:

FINE BERRIES

The finest strawberries that we have seen this season was a box left at The Capital office by our old friend, Mr. Joseph Beardmore, near Camp Parole. We placed them in our front window and they were admired by every passerby. They were as large as an English walnut, and 23 of the berries filled a heaping quart box. Mr. Beardmore has a small patch of three quarters of an acre from which he picks upon an average of 50 quarts a day. We don't know the variety, but we can vouch for the flavor as the most delicious fruit we ever tasted. If there is another grower in Anne Arundel that can excel this we would like to hear from him.

The year 1903 was an anomaly in the production of strawberries as rains caused all the strawberries to ripen at once, which drove the prices to record lows. This caused farmers to suffer great losses.

From ***The Baltimore Sun***, May 30, 1903:

BERRIES HALF CENT A BOX

Heavy Receipts Cause A Glut
In Local Market.

CHERRIES, TOO VERY CHEAP
Wharves Crowded With Boxes And Crates from
Maryland Counties - Peas Are Plentiful

Perhaps all previous records for strawberry prices were broken yesterday, when inordinately heavy receipts of this luscious fruit at Light Street and other wharves forced down the wholesale quotations to a half-cent a box, this being the figure at which a large consignment brought up by the steamer Tangler was disposed of. Offers at one cent a quart were common, and even at this nominal price fine berries went begging, the packers and other large consumers being simply swamped with the arrivals, which were swelled to extraordinary proportions by the rapid maturing of the crop in consequence of the recent frequent rains.

At a half-cent a box the farmer or the dealer who bought the grower's prospects sustains a loss of not less than 3 cents on every box, for the pickers get 2 cents a quart, the box and crating cost 1 cent more and the freight charge and commission amount to half-cent additional, which brings the total cost delivered in the market here to 3½ cents a box. At this rate it will, of course, not pay farmers to gather their crops or commission men to handle them, and the crop will be allowed to spoil on the vines. Meanwhile, however, as long as the present glut lasts, the thrifty housewife who delights in putting up preserves will be in her element, for she can secure all the berries she wants almost for the asking, the chief expense of "putting up" the fruit being for sugar and jars.

Proprietors of restaurants and other places where strawberries are served with ice-cream will reap a harvest, for it is not reported that the drop in the wholesale quotations has been so far reflected in the charges on bills of fare.

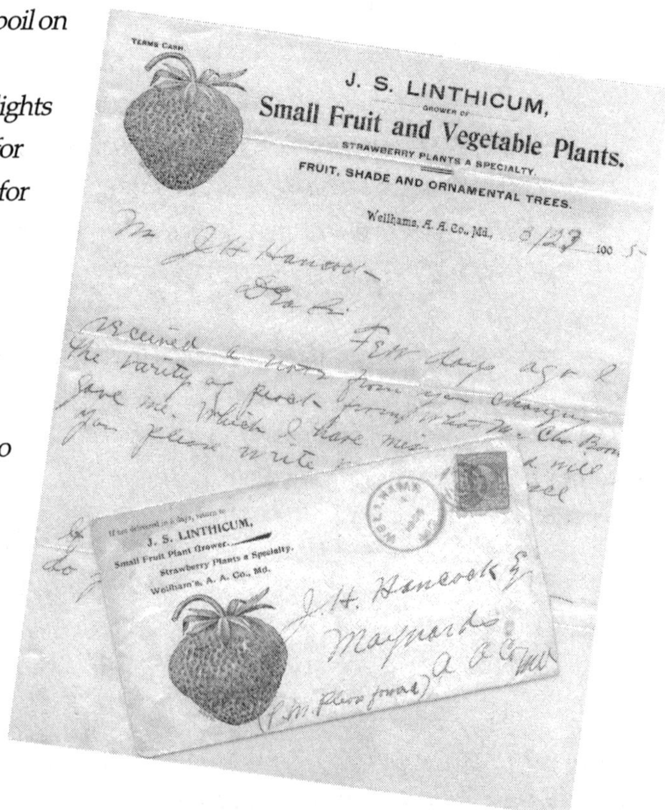

Letter and envelope of J. S. Linthicum reflecting the strawberry culture. (Collection of author).

From ***The Evening Capital***, May 29, 1913:

"SOME" STRAWBERRIES!

Variety Raised by Farmer Chas. D. Ridout

Anne Arundel strawberries are famous the world over. For inscipuaness(sic.), sweetness and fragrance there is no berry like the Anne Arundel berry, but seldom do they reach such a degree of size and quality as the strawberries raised on the farm of Mr. Charles D. Ridout, near Eastport.

Mr. Ridout has acquired the science of farming years ago and (as) a practical up-to-date farmer he has no peers.

In a box of strawberries farmed on Mr. Charles D. Ridout's farm, and a sample of the variety he puts out, not a single one in the quart measured less than 4½ inches longitudinally, and 4¼ inches latitudinally. Can you beat that?

During World War I, the truck farmers planted more vegetables to support the war effort and reduced their acreage in strawberries.

From ***The Evening Capital***, May 10, 1920:

SMALL ACREAGE CUTS CROP OF STRAWBERRIES

That strawberries will be scarce and high this season is the opinion of commission men in Baltimore. Most farmers are demanding from 25 cents to 30 cents a quart for their crops, boosting the retail price to 35 or 40 cents. The high price is due to the shortage caused by the berry crop being substituted by more profitable products.

Commission men of Baltimore City have purchased most of the State crop at the above prices. The crops are much smaller than in previous years and unless the berries can be obtained from another source the local markets will suffer. Due to the big prices and scarcity of strawberries, a number of dealers have refused to handle them. Other dealers admit they are taking a chance by purchasing the crops at the high price.

Stanley Steponawicz, an innovative Polish immigrant, found a good or maybe not so good "alternative" use for Anne Arundel strawberries during the prohibition years.

From ***The Evening Capital***, June 15, 1920:

ANNE ARUNDEL BERRIES FIGURE IN "STILL" CASE

Luscious Anne Arundel strawberries figured prominently yesterday in the case of a man, Moonshine and still, found in a raid on a room on Sharp Street, Baltimore, in the heart of the wholesale district, that city, made by Federal agents and the police.

Federal internal revenue agents and policemen of the Central District, Baltimore, visited the lodgings of Stanley Steponawicz, 45 years old, on the second floor of 126 Sharp Street, Baltimore, last night and found Stanley sitting at a table, capping strawberries.

"What are you going to do with the berries?" Sergeant Himmelman and Revenue Agent L. N. Scherf asked Stanley.

"Make wine, good wine," answered the native of Poland, still clipping the caps from the luscious Anne Arundel fruit.

Patrolman Mannus, of the Central District, moved close to Stanley's side, while the revenue man and Sergeant Himmelman sniffed the pungent odor of the room.

"Where's the schnapps?" asked Himmelman.

"Got no schnapps," replied the native of Poland, still capping the berries.

Agent Scherf read a search warrant to Steponawicz, but he paid little attention to it. Within a few minutes the police found a 20-gallon copper still, 18 quarts of home-made whiskey and 10 demijohns of whiskey.

Old-fashioned fireplaces built nearly a century ago were used as caches for the still and whiskey. A week ago Round Sergeant Burns received information that a still was in operation in the wholesale district. Night watchmen frequently detected the odor of the boiling mash and an investigation followed.

From ***The Evening Capital***, May 17, 1922:

FIRST STRAWBERRIES OF COUNTY
RETAILED AT
25 CENTS QUART

First shipments of real Anne Arundel county strawberries reached the Annapolis market today. They sold at the wholesale price of 18 and 20 cents per quart, and were retailed at 25 cents. The consignments that came in were from farms in the West River section, and were large, luscious and sweet. Berries from the Virginias and Carolinas have been on the market for several weeks, and have retailed at prices ranging from 20 to 40 cents per quart.

Almost simultaneous with the arrival of county berries here, fruit from county farms also reached the Baltimore markets and brought practically the same prices as prevailed here today.

Note that these strawberries came from the West River area in Southern Anne Arundel County.

From ***The Evening Capital***, May 18, 1922:

STRAWBERRY TIME

Appearance of Anne Arundel strawberries on the market is hailed with delight by discriminating epicures, for it is a well known fact that the strawberries grown in this county are about the last word in the fruit line.

Delectable, succulent and possessed of a sweetness the like of which cannot be found in any of the berries grown beyond our confines, the luscious Anne Arundel product has gained for itself an enviable reputation both at home and in outside realms.

As a matter of fact, in the adjacent cities every householder knows that super excellent strawberries really mean the Anne Arundel kind of fruit; so much so that our own product has gained fame and taken precedence over all other varieties.

The soil of Anne Arundel is especially adapted to the growing of deluxe strawberries, as well as other kinds of high class farm products, and our farmers are painstaking in their efforts to see that the produce put on the market is kept at high standard. Judging from the strawberries which already have reached here, the excellent standard set in former years is being maintained.

By 1922, southern berries transported in refrigerated freight cars were beating the Anne Arundel crop to the local markets. The first produce to reach the market commanded the highest price which represented major profits for the farmers. During the 1920s, roads were improved and trucks were built to travel long distances with lights powerful enough to travel at night. By the late 1920s, fruits and vegetables were brought to Baltimore from the South in refrigerated trucks long before Anne Arundel produce could ripen. Anne Arundel farmers were losing their high profits, which eventually became one of the major factors in the demise of the Anne Arundel truck farm. With the development of Anne Arundel land, by the late 1920s Eastern Shore counties surpassed the strawberry production of Anne Arundel County. The Anne Arundel strawberry continued its slow decline throughout the 1930s. On May 18, 1940, ***The Evening Capital*** published the following editorial which could be considered an obituary for the Anne Arundel strawberry:

STRAWBERRY

Once upon a time, people were wont to talk about the strawberry season and to look forward to it with delightful expectation. It brought visions of strawberry shortcake with mashed berries between the stories of biscuit, contrasting with the delicate tints of rich cream. And there was the social angle, the strawberry festival which brought together the elite of the neighborhood and afforded youthful gallants the opportunity of financing a treat for fair ladies.

Gone, gone, are these amenities, sacrificed beneath the juggernaut wheels of advancing science and victimized by the wealth, prosperity and climatic resources of this great country. Everywhere else in the world seasons for strawberries still survive. In Europe, where national boundaries are close together and national self-sufficiency is a coddled ideal, seasons for strawberries are well-defined and short.

But in this country good ripe, fresh strawberries at reasonable prices are to be had in the depths of winter, and long before the strawberry plants have blossomed in Anne Arundel gardens, strawberry shortcake is a commonplace for all who have not tired of it. All this has been brought about through the wide expanse of Uncle Sam's territory, and through progress of science in horticulture, refrigeration and transportation. Good old strawberry! Long may she wave!

Vivian, George and Margaret Schmidt packing strawberries on their Pasadena farm around 1933. (Photo courtesy of Marion Warren).

Other Produce

While strawberries were the most exotic crop from the truck farms, vegetables and other fruits constituted a majority of the crops grown on the Anne Arundel farms. The season started with strawberries, followed with peas, beans, raspberries, tomatoes, corn, potatoes, sweet potatoes, peppers, cucumbers, carrots, radishes, cabbage, lettuce, asparagus, beets, squash, melons, pumpkins and orchard fruits such as peaches, pears, apples, and cherries. While none of these had the impact on the culture that strawberries had, they all carried Anne Arundel's reputation for superior produce. Tomatoes and melons were especially in demand because of their "Anne Arundel" quality.

John Dick Shipley watching men load "Long John" cantaloupes onto his 1928 Model AA Ford truck. John Shipley originated the "Long John" cantaloupe on his farm near Bayside Beach. (Photo courtesy of Jack Kelbaugh).

The Griffith and Turner seed catalogue for 1920 states that Anne Arundel County "is justly celebrated for growing the finest cantaloupes in the country. We almost might say the finest in the world." This reputation rested mainly on several varieties of the green-meat melons which were developed by farmers in Anne Arundel County. The 1910 catalog from The Meyer Seed Company in Baltimore advertised seed for the "Long John," the "Bottomley," and the "Knight," all cantaloupes described as "true products of Anne Arundel County which is famous for its fine cantaloupes." Their catalog of 1930 still offered the same varieties of Anne Arundel cantaloupes.

"Long John" cantaloupes, one of the green-meat melons developed in Anne Arundel County. (Photo courtesy of Jack Kelbaugh).

The Bottomley was developed by Robert Bottomley on his farm on Fort Smallwood Road. A 1928 seed catalog described the Bottomley as "Beautifully netted with deep green ribs, giving it a most attractive appearance. Thick meated, very crisp and fine flavor. Handsome greenish color shading to a frosty salmon next to seed cavity...a wonderful shipper. Retains its flavor and firmness longer than any other melon. A great favorite wherever grown."[21] The Long John, developed by John Dick Shipley, and the Knight, developed by Adam Neidert, along with the Bottomley, dominated the Baltimore market for decades.

In 1917 or 1918, Louis Schramm developed a variety of the Bottomley which he called the "Schramm." By the early 1930s, the Schramm was very popular in Virginia. The Schramms continued to grow their melons until they gave up their farm in the late 1990s. The Long John is still grown in Anne Arundel County by Gordon Wirth of Pasadena, Melvin Upton, a third generation truck farmer, and Jimmy Schillinger, both on New Cut Road. All three market their large football shaped Long Johns in the Farmers' Market on Riva Road.[22]

Again the contemporary newspapers reflect the truck culture:

From ***The Maryland Republican & Advertiser***, August 20, 1870:

> ***OUR MARKET*** *yesterday morning was more beautifully supplied with all kinds of vegetables and fruits than we have seen it for some time. The number of persons from the county bringing in produce was quite large. The whole space in the interior of the market square, benches and pavements, were filled with the "kindly fruits of the earth," and long lines of carts and wagons extended along on each side of the market. Watermelons and cantaloupes, peaches, apples and pears were in profusion, and sold at different prices, according to size and quality.*

From ***The Maryland Republican & Advertiser***, April 1, 1871:

> ***Annapolis, April 1, 1871.*** *...A greater portion of Anne Arundel county might be termed the garden county of Maryland, since so many of its farmers have turned their attention to the cultivation of fruits and vegetables so extensively, as to exclude almost entirely the raising of wheat, corn and tobacco, except for their immediate use. The trade with Baltimore and Philadelphia, for the produce raised in this county, has become so extensive along Rock and Bodkin Creeks, as to necessitate the placing of a steamer on the above creeks, instead of the slow sailing crafts that were formerly used for the transportation of marketing. In order to meet these demands, Mr. John H. Barnes, of Baltimore, has had built, and put upon the Bodkin Creek section of our county, a steamer suitable in every respect for the quick transportation of produce to market. This arrangement has met with such success the past season, and so great has been the demand for freight, to and from the city, that Mr. B. we are informed, contemplates placing a similar steamer on Rock Creek the coming season, in order to supply the wants of that section of our county.*

Top: Truck filled with bags of tin cans used by Charles Hancock Calvert. Cans were used in the hot-beds to grow cantaloupes and tomatoes from seeds. (Photo courtesy of James B. Calvert).
Middle: Hot beds of Charles Hancock Calvert. (Photo courtesy of James B. Calvert).
Bottom: Truck of Gilbert H. Clark loaded with cantaloupes. (Photo courtesy of Ora and John Smith).

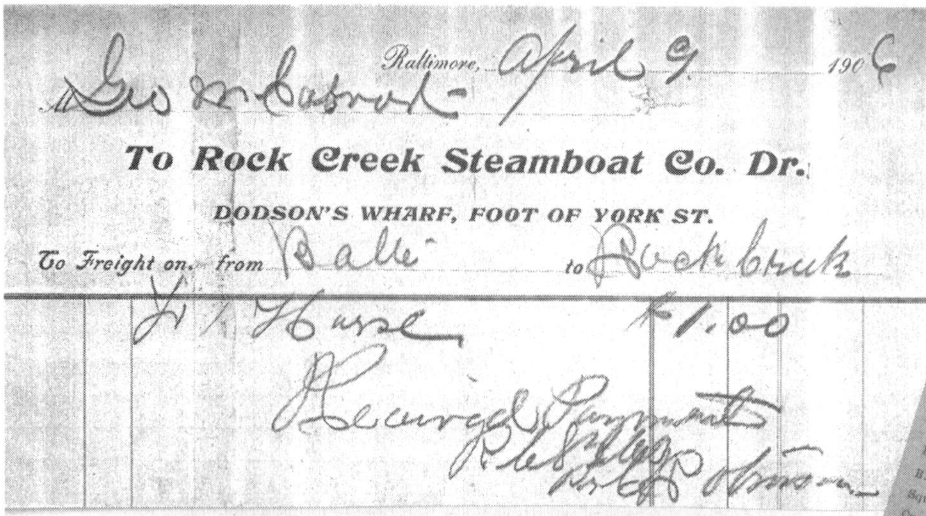

Baltimore, April 9 1906

To Rock Creek Steamboat Co. Dr.

DODSON'S WHARF, FOOT OF YORK ST.

To Freight on from Balto to Rock Creek

1 Horse $1.00

Received Payment

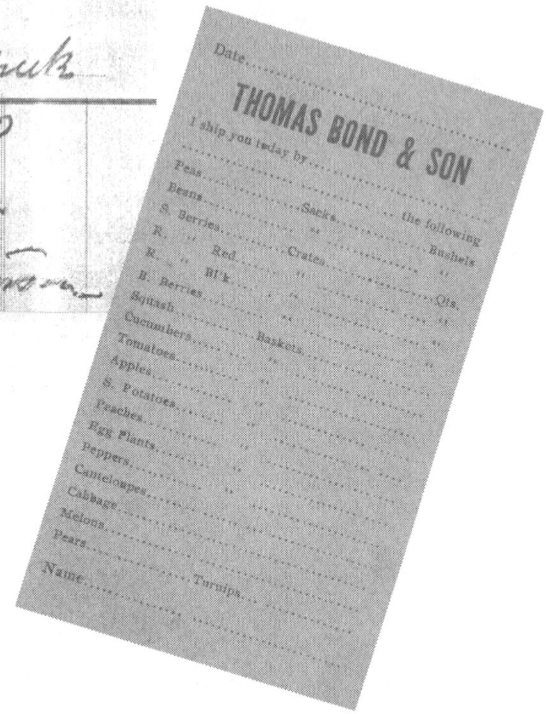

Date.....

THOMAS BOND & SON

I ship you today by..... the following

Peas..... Sacks..... Bushels

Beans..... " ..... "

S. Berries..... Crates..... "

R. " Red..... " ..... Qts.

R. " Bl'k..... " ..... "

B. Berries..... " ..... "

Squash..... Baskets..... "

Cucumbers..... " ..... "

Tomatoes..... " ..... "

Apples..... " ..... "

S. Potatoes..... " ..... "

Peaches..... " ..... "

Egg Plants..... " ..... "

Peppers..... " ..... "

Canteloupes..... " ..... "

Cabbage.....

Melons.....

Pears.....

Name..... Turnips.....

Top: Steamboat receipt for shipping from Baltimore to Rock Creek, April 9, 1906. (Courtesy of James B. Calvert).
Right: Form used for shipping produce to Baltimore via steamboat. (Courtesy of James B. Calvert).

From ***The Maryland Republican & Advertiser***, May 13, 1871:

> Under Local Correspondence. Second District:
>
> *The culture of peas on a large scale has been undertaken by the Messrs. Wellham, near Crownsville. Their land is well adapted to this leguminous delicacy, the proper accompaniment of lamb. Lamb, peas and strawberries! How the mouths of the votaries of Epicures water at the very thought of these delicacies! Only think, sixty acres in peas! the size of an ordinary farm, and this planted in peas alone, besides strawberries, peaches, melons, tomatoes, cultivated by these enterprising and energetic gentlemen, who understand their culture to perfection.*

From ***The Maryland Republican & Advertiser***, June 3, 1876:

> ***Fine Asparagus.*** *- We received from the farm of Mr. Thomas C. Boone, on Saturday, two bunches of the finest specimens of asparagus ever brought to this market. Mr. Boone is a thrifty farmer and truck grower on North Severn, in this county, and has supplied our market with some of the finest vegetables grown in this section. He is au fait in matters pertaining to horticulture.*

Walter D. Klingelhoefer operated a farm on Mountain Road in Pasadena. This farm was the only truck farm in Anne Arundel County to plant over 50 acres in asparagus. He built a special asparagus packing shed where the asparagus was trimmed and tied in bunches for market.

Shed used to prepare asparagus for market. Walter Klingelhoefer farm, Mountain Road. (Photo courtesy of Dolores Klingelhoefer.)

From ***The Maryland Republican & Advertiser***, May 3, 1879:

> ***Peach Crop.*** *- Everything so far indicates that the peach crop in this county will be unprecedented this year, both as to quality and quantity. In all parts of the county the most flattering prospects exist, and information from every district is favorable.*
>
> *Mr. R. T. Brice, North Severn, who has seven thousand peach trees on his farm, informs us that he has never seen his trees so full of buds before, and seldom seen them presented a healthier appearance.*

From ***The Evening Capital***, September 2, 1887:

FINE PEACHES

> *Mr. A. A. Stinchcomb of the Third District is fortunate in having a crop of five hundred or six hundred boxes of peaches. All that he has sold so far have brought him $2 a box. He says he has an orchard of 14000 trees, and that it is the prettiest peach orchard that ever grew out of the ground.*

Alfred Asbury Stinchcomb had one of the largest and finest peach orchards in the county, covering three hundred acres. His farm was located on the west shore of the Little Magothy River, which is now a part of Cape Saint Claire. He constructed two large 40-foot long row boats which he used to ferry peaches to a pungy boat anchored in deep water. He built two special "peach wagons" which he loaded with bushels of peaches, and then ran both of them into the water axle deep. He would position the large row boats between the two wagons and transfer the peaches into the row boats. The row boats would then carry the peaches to the larger pungy boat which would carry the peaches to the Baltimore markets.[23]

Peach growing in Maryland had its origin in Anne Arundel County. The first large peach orchard was planted in the early 1800s by Thomas Robinson with over 18,000 trees. They were used primarily for making peach brandy. By 1810 orchards were planted a few miles south of Baltimore by Richard Cromwell, Abner Linthicum, and Ellis Thomas, whose farm was located at the present site of Cedar Hill Cemetery. By mid-century, the demand for peaches greatly increased as the peach was found to be an ideal fruit for canning. To meet this demand, more peach orchards were planted throughout the county. In 1850, Mr. Wilson planted peaches on the Magothy River.[24] By 1909, Anne Arundel County had over 188,000 peach trees with an annual production of over 27,000 bushels.[25]

Right: Charles Shipley washing tomatoes with daughter Amy. (Photo courtesy of Jack Kelbaugh and Amy Shipley Mahanes).
Below: Alfred Asbury Stinchcomb's special wagons used for loading peaches into a large row boat to transfer the peaches to a "pungy" boat in deeper water, 1911. (Photo courtesy of Sara Anne Stinchcomb).

From ***The Evening Capital***, May 29, 1890:

ACTIVITY IN THE TRUCK BUSINESS

The produce wharves in Baltimore are beginning to show activity in the truck business from the farms of Anne Arundel county. Bowly's wharf yesterday was piled with Wakefield cabbages, which when unloaded from bugeyes, sloops and pungies were quickly trimmed and boxed for shipment. Philadelphia was prominent in the orders being filled, but large quantities were shipped West.

From ***The Evening Capital***, May 25, 1903:

WORK ON CANNING FACTORY

Thomas M. Lloyd, contractor and builder, of Annapolis, broke ground on Saturday for a canning factory for W. W. Roberts at Camp Parole. The factory will be directly opposite the station. Mr Lloyd will also build a warehouse, store and dwelling for Mr. J. F. Barrow at Camp Parole, for which ground has been broken and work was begun today.

Tomatoes were especially suitable for canning and grew like weeds in Anne Arundel soil. An Anne Arundel tomato was so sweet that you had no trouble believing it was a fruit. Many canneries specialized in tomatoes and, by 1890, Anne Arundel County grew almost 25 percent of the total U. S. production.[26]

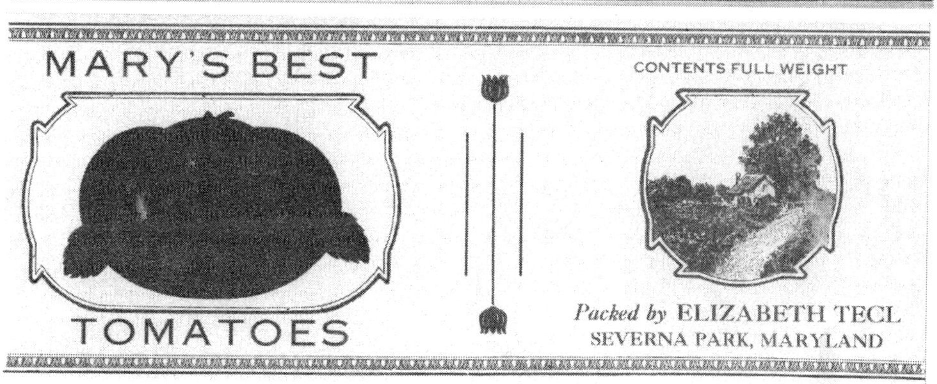

Top: Can label from J.C. Schultz, Anne Arundel Pride, Pasadena, MD. (Author's collection).
Bottom: Can label from Elizabeth Tecl, Mary's Best Tomatoes, Severna Park, MD. (Author's collection).

The Markets

Truck farming could not exist without easy access to markets. In the early days, only farmers living within a few miles of Baltimore or Annapolis could transport their produce to the city markets in time for the early morning markets. As the population of Baltimore increased, more farmers resorted to raising truck, and as the roads improved and bridges were built, truck farming expanded to greater distances from the population centers.

Many farmers depended upon sailing vessels to carry their produce to the markets. In 1840, the Annapolis-Elkridge Railroad was opened through the northwestern part of the county which opened vast lands to truck farming. More land was opened to truck farming with the addition of the Baltimore and Potomac Railroad in 1872. With the waterways on the east and the railroad in the west, hardly an acre of land in northern Anne Arundel County was more than ten miles from transportation to a market. The railroads also opened the Washington market and more distant markets to the north and west. The steamboat expanded the truck farming area to an even greater distance, enabling the farmers with access to waterways in the southern part of the county to grow truck produce.

The growth of canneries also expanded the markets for certain truck produce. Baltimore became the canning center of the United States by canning oysters in the winter and fruits and vegetables during the rest of the year. By 1900, it is estimated that there were over twenty canneries operating throughout the county, mostly along the railroads. The canneries provided greater markets for the truck farmers and production increased to satisfy the new demands.[27]

Top: A. Pumphrey, Pride of Benfield Tomato can label. (Author's collection).
Above: Marsh Market in the 1920s (Photo courtesy of The Baltimore Sun*).*

The main markets for Anne Arundel truck produce were Baltimore, Washington, and Annapolis, but crops were shipped as far north as Toronto and Montreal, Canada, and as far west as Denver. Baltimore was the major market for domestic sales as well as for shipment to other cities. [28]

Baltimore had several wholesale markets, the main one being the Centre or "Marsh" Market, built in 1787 on a drained swamp, hence the name "Marsh" often called "Mash" Market. In 1883, a section was devoted to dealing wholesale in farmer's produce. An ordinance of the City Council allocated that part of Centre Market lying between Lombard and Pratt

Streets as a "wholesale market for country produce brought into the city in wagons and carts." A wholesale produce section of the market was built in 1907 after the original market was destroyed in the great Baltimore fire of 1904. The produce section was designed to hold two trucks abreast, accommodating about 250 vehicles. Traffic congestion became a major problem as the trucking industry rapidly increased. Other Baltimore markets included the Pennsylvania Railroad Produce Terminal, the Baltimore and Ohio Railroad Camden Street Yards, the Producer Piers, and the Camden Street Produce District (located at intersection of Camden and South Charles Streets.)[29]

Farmers who shipped their produce to Baltimore by boat usually used Long Dock, also on Pratt Street. The docks were congested, especially during watermelon season. Both sides of Pier 4, which made up Long Dock, were filled with craft. Some were so heavily laden with watermelons that their decks were awash. The boats were three or four deep waiting for a berth at the dock. There were schooners, bugeyes and skipjacks along with the steamers.

Above: Anne Arundel watermelons. (Photo courtesy of Jack Kelbaugh). Left: Boats unloading watermelons at Baltimore's Long Dock. (Photo courtesy of Jack Kelbaugh).

The scene was akin to a carnival. Vendors of ice cream, snowballs and hotdogs made their stand. Hucksters' wagons backed up to the boats, and the area was a forest of masts. Young and old sat on the curbing of the pier eating watermelons purchased from the boats for a pittance. The water was a mess, covered with rinds, whole melons and rotten ones. Young boys would target the floating melons with home made spears.

Eventually power boats replaced the sail boats and the scene was never the same. Large trailer trucks soon monopolized the trade and an era came to an end at Long Dock.[30]

Farmers transported their produce to the markets and sold their crops directly to a commission merchant or to a wholesale receiver who in turn passed the produce on to commission merchants or brokers who served as agents for other markets or large quantity orders of produce. The commission merchants or wholesale receivers also sold produce to "jobbers" who passed the crops onto street peddlers or hucksters, commonly called "a-rabs." The a-rabs became legendary in the streets of Baltimore and other cities as they peddled the produce from hand or horse-or mule-drawn wagons and are still seen in the streets of Baltimore. Their cry of "Annarannel" melons or strawberries became famous in the streets of Baltimore and in many streets of the other cities where Anne Arundel produce was sold. The following contemporary tale illustrates the great reputation of Anne Arundel County truck produce:

A black "a-rab" was pushing his cart through the streets of Baltimore yelling in a singing voice as he moved slowly:

"Annarannel Berries...
Annarannel Peas...
Annarannel Beans...
Annarannel Oranges..."

An observer flinched as he heard "Oranges!"

He stopped the vendor and questioned, "Did I hear you say, 'Anne Arundel Oranges?' Any fool knows they don't grow oranges in Anne Arundel County..."

"I know boss," replied the a-rab, "But that's the only way I can sell 'em."[31]

Modern A-rab with horse and cart. (Photo courtesy of The Baltimore Sun*).*

With the decline of truck farming in Anne Arundel County during the 1930s, Marsh Market began to receive trailer loads of produce from the South and other truck farming areas in the East. By the late 1950s it was nearly impossible to do business in the old facility built for horse-drawn wagons. The city replaced the market in 1960 with a new market at 6400 Pulaski Highway. The old downtown markets were purchased by the city for the Inner Harbor development. The wholesale markets now exist as the Maryland Food Center at Jessup, which began operations in January 1976. Managed under state authority, the Jessup facility has been dubbed the Produce "Wall Street" of the East.[32]

Roadside produce stand, Anne Arundel County, 1950s. (Photo courtesy of Maryland Department of Education).

The other produce outlived the strawberry in Anne Arundel County and was marketed by many farmers on roadside stands. During the late 1940s and the 50s, over 200 roadside produce stands were operating in Anne Arundel County by farmers who no longer marketed all their produce in the Baltimore markets. Some stands were owned by merchants who purchased the produce from local farmers or the Baltimore wholesale market. Today the few farmers left and some part-time farmers sell their produce in several Farmers' Markets; yet a few roadside produce stands are still found scattered throughout the county, remnants of the glory days of the Anne Arundel truck farm.[33] The Schillinger's Vegetable Farm (Papa John's) is an exception as they market much of their produce to local food markets, operate a large roadside stand, and sell produce in the Farmers' Markets.

Right: 90-year-old Melvin Upton, third generation truck farmer at Farmers' Market on Riva Road. (Photo by author). Below: Schillinger's Farm Produce (Papa Johns) at Riva Road Farmers' Market. (Photo by author).

Matthew Weller and his wife, Carol, run a roadside produce stand near the Annapolis Mall on Generals Highway. Matt sells corn and melons which he obtains from the Eastern Shore, but still sells tomatoes and peaches grown by farmers in Anne Arundel County.

Residents in the Severn area still enjoy fresh farm produce from the roadside stand of Ridgely Boyer. On the Pasadena peninsula, where truck farming once flourished, only three farmers still grow truck produce in any quantity: the Schramms, Gorden Wirth and Rick and Cindy Holt. They sell their produce from roadside stands or in the farmers' market.[34]

Matthew Weller, farm produce outlet on Generals Highway. (Photo by author).

CHAPTER THREE NOTES

1. Fletcher, *The Strawberry in North America*, p. 2.
2. Wilhelm, Stephen and Sagen, James E., *A History of the Strawberry: From Ancient Gardens to Modern Markets* (Berkeley, Ca.: University of California, Division of Agricultural Sciences, 1974), pp. 6-11.
3. Fletcher, *The Strawberry in North America*, pp. 6-12.
4. Ibid, p. 17.
5. Burton, R. Lee, Jr., *Canneries of The Eastern Shore* (Tidewater Publishers, 1986), p. 130.
6. Fletcher, S.W., *Strawberry in North America*, pp. 29,44.
7. Fletcher, S.W., *Strawberry Growing*, New York, The MacMillan Company, 1917, (preface).
8. Fletcher, S.W., *Strawberry in North America*, p. 30.
9. *Gardener's Monthly*, 1859, p. 106.
10. Fletcher, S.W., *Strawberry in North America*, p. 42.
11. *The Baltimore American*, May 22, 1862, "Strawberries and Peas."
12. Fletcher, S.W., *Strawberry in North America*, p. 43.
13. Ibid, p. 63.
14. Ibid, pp. 308-312.
15. Ibid, p. 42.
16. *The Evening Capital*, May 27, 1910, "Repeated Tonight."
17. Numerous articles and ads appear during strawberry season (late May through June) in *The Evening Capital* and *The Sun*. Strawberry festivals are still held in Anne Arundel County reflecting the old strawberry culture.
18. *The Baltimore News*, May 18, 1913, "Pastor Condems Church Festivals."
19. *The Evening Capital*, May 25, 1942, "Strawberry Festival Cancelled-No Sugar."
20. Fletcher, S. W., *Strawberry Growing*, pp. 265-267.
21. *The J. Bolgiano Seed Company*, Baltimore, MD, 1928.
22. Cunningham, Isabel S., "Anne Arundel's Famous Green-Meat Cantaloupes." *Anne Arundel County Notes*, 27, no. 4 (July 1996): 3,4,8-10
23. Masonetta May Stinchcomb Waring related this story in an unpublished 1953 document about her parents, Alfred and Sarah Stinchcomb.
24. Gould, H. P., *Maryland Agricultural Experiment Station, Bulletin 72*, March 1901, "Peach Growing in Maryland," pp. 128-131.
25. Census of U. S. 1910 with Maryland Supplement, p. 616.
26. Maryland Agricultural Experiment Station, Bulletins 1-5, Annual Reports 1888-1891. Published by the Experimental Station at the Maryland Agricultural College, P. G. County, Md. Annual Report of Trustees for 1889, p. 5.
27. Riley, Elihu S., *A History of Anne Arundel County, in Maryland* (Annapolis: Charles Feldman, 1905), P. 122. Riley lists the following canning factories or fruit packers: Martin Wagner Packing Co.; Noah H. Green, Benfield; David Wigley, Gambrills; Gambrill's Packing Co.; W. L. Gardner & Sons, Odenton; C. G. Summers, Odenton; George M. Murray, Odenton; K. Boswell, Waterbury; C. Nocklitz, St. Margarets; Richard H. Maynard, Woodwardsville; O. P. Roberts, Chalk Point & Galesville; Charles S. Tate, Arnold. We add to this list: J. C. Schultz, Pasadena; Elizabeth Tecl, Severna Park; A. Pumphrey, Benfield.
28. *The Sun*, June 13, 1910. Anne Arundel produce was shipped to Chicago, Denver, New York, Rochester, Buffalo, Boston and Canadian cities.
29. Burdette, R.F. & DeVault, S.H., *The Baltimore Wholesale Fruit and Vegetable Markets* (University of Maryland, Agr. Expt. Sta. Bul. No. 409, May 1937).
30. Burgess, Robert H., *The Sunday Sun Magazine*, Sept. 12, 1954. "I Remember...Watermelon Boats at Long Dock."
31. This tale was told by the late Bessie Zeigler from Harendale, MD. Her husband was a small farmer in the Pasadena area.
32. *The Evening Sun*, July 24, 1985, "Glory Days at Marsh Market" by James Miller Merritt.
33. The Maryland Farmers' Market Directory for 1999 lists four markets for Anne Arundel County: Anne Arundel County Farmers' Market on Riva Road and Harry S. Truman Parkway; Crofton Farmers' Market at Village Green (North); Piney Orchard Farmers' Market at Stream Valley Road (Off Route 170) Visitor's Center Lot in Odenton; Jones Station Farmers' Market on Ritchie Highway (Route 2) & Jones Station Road in Severna Park. Maryland Department of Agriculture web site address: **http://www.mda.state.md.us/market/fmd.htm**
34. Cunningham, Isabel S., *Between Two Rivers*, p. 119.

The Pickers

Strawberry Time

Strawberry time in the fields of home
Strawberries under the clover foam;
Down to Arundel the pickers go,
Women and children from row to row
Dew of the morning, and birds in rhyme,
Chattering, chattering, strawberry time!

Folger McKinsey

In the early days of truck farming, before the Civil War, most farmers' families and regular farm hands did the bulk of the work. Some farmers had a few slaves to help during the picking season, but if more labor was required, local residents were hired to meet the temporary demands of the harvest. Many early truck farmers were still involved in growing tobacco which created an even greater demand for extra labor. As the truck farms grew in size, finding extra labor became a special problem for the farmers.

During the Civil War, farm labor became scarce as men, young and old, left the farms to serve in both the Union and the Confederate armies. The war also created a large increase in the demand for truck produce of all kinds. Fresh vegetables and canned foods were especially needed in great quantities to feed the large army. Thousands of Union troops flooded the nation's capital just a short train ride from Anne Arundel County. The railroads from Baltimore and Annapolis to Washington were guarded by detachments of Union soldiers to keep rebel troops and Southern sympathizers from sabotaging the rails and to repair the rails if they were sabotaged. The railroads were strategically vital as they served as one of the major logistical lines to support the war, linking the North to Washington. To avoid the Southern hotbed of Baltimore, Union troops were shipped to Annapolis from Havre de Grace, and were transported into Washington from Annapolis, using the Annapolis and Elkridge Railroad.

The troops guarding the railroad and those stationed in the area needed to be fed, and the local farmers were tasked to produce more food. As the war progressed, Camp Parole was established just west of Annapolis, along the railroad. At its peak, Camp Parole housed over 20,000 soldiers. In addition, several thousand more troops were stationed in the medical and logistics camps in Annapolis. The food supply was critical.

The farmers were called upon to increase their production of food products. Prices for truck produce increased with the demand, and farmers resorted to growing more vegetables and less tobacco. Increased production created a need for more labor, especially during the picking season. Even those farmers with slaves needed additional labor to meet the demands of the harvest.

The labor shortage stifled the rapid expansion of truck produce during the war years. This problem intensified in late 1864 when a new Maryland constitution based on the Emancipation Proclamation freed all the slaves in the state. Many of the freed blacks fled the farms for the cities where industrial labor markets paid higher wages. ***The Baltimore Sun*** reported on June 24, 1863, that "the farmers in this county, owing to the scarcity of labor, have applied to the colonel commanding at Camp Parole for paroled prisoners to assist them in harvesting their crops." In a letter to her cousin in Newport, R.I., Susan Brice, widow of Charles Carroll Brice, owner of the Pendennis Mount Plantation on the north shore of the Severn, wrote: "We have a parole soldier living with us. He is an excellent hand and understands (thoroughly) all farm work, so willing and obliging. We shall miss him very much when he leaves. He has been with us a month...."[1]

Despite the labor shortage, Anne Arundel farmers produced more truck crops than ever before and many continued after the war to help supply the growing needs of the rapidly increasing Baltimore population and the growing needs of the canneries. Truck farming soon monopolized the northern part of the county. But the labor shortage continued after the war as returning soldiers found better jobs in the growing urban industrial community.

In the June 18, 1871, edition of the ***Republican and Maryland State Capital,*** a letter to the editor from a farmer in the third District in Anne Arundel County said:

> *Crops are on the whole looking remarkably well. There seems in this section to be quite a scarcity of labor, and farmers are somewhat in the grass, in consequence.*
>
> *If some "agricultural physician" could doctor this labor question to the extent of changing its whole appearance and working, he might confer something like immortality upon himself, or at least desire to be forever held in grateful remembrance by Maryland farmers. It is very evident, that unless something is done to secure intelligent, reliable and permanent labor, the agricultural interests must languish. It is about time to cease expecting farmers to put their own shoulders to the wheel. They are quite willing generally to do the praying, but some one else must do the work of securing labor....*

One year later on June 15, 1872, the same newspaper ran the following news item:

> ***German Labor.*** *- Mr. James Legg, of our city (Annapolis) has secured, through the emigrant agent at Baltimore, some thirty German laborers for his truck farm, in this county, to which place they were sent and assigned to their several duties on Tuesday last. Mr. Legg, we learn, is much pleased with their mode of working, and pronounces them an industrious and thrifty set of people.*

From this account it appears that in 1871 there was no great quantity of immigrant laborers working on the Anne Arundel truck farms. Mr. Legg in 1872 seems to have laid claim to that "agricultural physician" to "doctor" the labor problems with immigrant workers from Baltimore. Yet Mr. Legg has far from achieved the immortal status which was forecasted.

The farmers wasted no time in filling their fields with immigrant workers. The very next season *The Maryland Republican* reported on June 7, 1873, that "the importation of the German element into our county, just at this time, far exceeds the foreign arrivals at New York. Our strawberry fields are perfectly alive with our Teutonic cousins, of all ages and sex, from the little urchin that crawls around, to the gray-haired sire and matron—all busy as bees plucking this delicious fruit for market."

The immigrants were a natural source to serve as part-time pickers for the truck farms. Most had come from rural farm communities in Central Europe and were still seeking the "better life" in America while living in urban ghettos. By 1870, there were over 80,000 foreign-born white citizens in Maryland and by 1910 they had increased to over 100,000.[2] Most of these lived in ghetto communities in Baltimore. They were looking for opportunities to breathe the fresh air of the country and at the same time earn extra cash, keeping alive the hope of someday owning their own farm. Many immigrants worked in the Baltimore canneries and some were employed in the great Baltimore market houses. Both of these areas provided possible contacts with farmers which may have resulted in the recruiting of immigrant families to migrate to the truck farms during the picking season.

The immigrants were the answer to the truck farmer's labor problems and the migration to the country farms was eagerly sought by the immigrants. The system became an annual affair and, at its peak, over 10,000 immigrants migrated to the Anne Arundel truck farms. [3]

Arrangements were made with a "rowboss," often a bilingual immigrant, who negotiated with the farmers and recruited the pickers. The farmers could tell almost to the hour when the strawberries would be ready for picking. He would contact the rowboss to let him know when the pickers would be needed. On the designated day the farmers sent their straw-filled wagons to Baltimore to pick up the pickers and head for the farms. Some traveled to the farms by railroad, while steamboats carried many pickers to the farms which were on the water.

The ***Annapolis Gazette*** of June 9, 1874, quotes the ***Baltimore American*** on the first parade of immigrants coming into the Anne Arundel fields.

> *Some curiosity was excited in the business portion of the city (Baltimore) yesterday by a number of two-horse wagons driving along Lombard and Baltimore streets, crowded with men, women and children, coming from the direction of Fell's Point. It was first supposed that they were foreign emigrants, just landed from some foreign ship, but as it turned out the party was simply on their way to the strawberry patches of Anne Arundel county to assist in gathering the crop now ripening there for market. These pickers work in gangs, and are paid so much per quart. They usually occupy the barns on the farms where they engage to work, and do their own cooking. They move from place to place, wherever their services are required, and when the strawberries are all gathered turn their attention to*

raspberries, blackberries and the like. Thousands of persons are required to gather the berry crops of Anne Arundel county alone. Whole families are engaged at this employment, and their habits during the summer months are not unlike those of the gypsies. In the fall nearly all of these people may be found at work in oyster and fruit-packing establishments of this city.

Above left: Off to the berry farms of Maryland. Fells Point, Baltimore, 1909. Above: A street full of Baltimore immigrants ready to start for the Anne Arundel County berry farms. Left: Picker's Shanty, Bottomley farm in Anne Arundel County, Maryland, 1909. (Photos by Lewis Hine. Courtesy of Prints and Photographs Division, Library of Congress).

Pickers described the event as an adventurous summer outing to the country. The children especially looked forward to their farm adventures. A small article in ***The Evening Capital*** of May 21, 1886, mentions the pickers:

STRAWBERRY PICKERS

The strawberry and pea pickers are flocking into the Third District, in large numbers. The business will be fully underway the first of next week, and large quantities of berries will be shipped to Baltimore by sailboats.

Whole families settle in the district, with their furniture and cooking utensils, finding shelter in barns and other outhouses belonging to the truck farmers. They make good wages and live happy, jolly, Bohemian lives during the season.

A few years later, the May 30, 1892, issue of ***The Evening Capital*** gives a very detailed account of the life of a picker on an Anne Arundel truck farm:

ARMY OF HARVESTERS

Life of the Strawberry Pickers In Anne Arundel.

THOUSANDS OF LIVELY WORKERS.

Sleeping in the Farm Barracks.Gypsy Evenings out of Doors. The Bright Check Girl. A Succession of Crops to Keep All Busy. The Dining Houses.

Thousands of men, women and children armed with cooking utensils and bed-clothing have just invaded Anne Arundel county. Here they will remain until the last vestige of the season's crop of berries, peas and beans have disappeared under their operations. The invasion began two or three days ago and still continues. Wagon load after wagon load of humanity has rolled over the Light Street and Harman's bridges bound for the strawberry patches, and the sloops, bugeyes, canoes and pungies that have sailed from the County Wharf, Baltimore, have arrived to the truck farms along the river and bay sides, big loads of passengers, all bent on the same errand, that of relieving the vines of their fruit.

The strawberry pickers are recruited from the neighborhoods about the packing houses in Baltimore, and they are of almost every nationality. Bohemians, Poles and Germans predominate, with a fair sprinkling of Americans, Italians and colored people. The work is eagerly sought and is regarded by many as a big frolic, lasting as it does for several weeks. It affords those who engage in it a rude sort of summer vacation, with opportunities for making money instead of spending it, and gives them relief from the city's heat and dust. Then, too, everybody is on precisely the same footing. Age and experience count for but little, and the urchin of eight, if his fingers be nimble, can earn as much in a day as his father, and if he be smart as well can fill himself with the luscious fruit. All these things contribute to the pleasure of the pickers, and on their way down they shout and sing like a crowd of school children on a picnic.

The strawberry season is just opening, and while very few barrels have been picked as yet, everything about the truck farms are in a bustle getting things in shape. As soon as the pickers arrive they are taken to the "barracks" and left to look out for themselves. The barracks where the pickers live while on the farms vary according to the means of the farmer and the size of his patch. The more pretentious ones are long, low two story buildings divided into small rooms, a family being put into a room, while the whole building is made to accommodate from 50 to 100 persons. More often they are simply old tenant houses. There is absolutely nothing in them when they are turned over to the pickers. They have to make their own household arrangements, the only things furnished by their employers being water and the wood, and straw for bedding. The only requirement of the grower is that the building shall shelter a certain number of people. The pickers settle all difficulties among themselves, and those first on the grounds get the best quarters.

The life is as near gypsy like as anything can be. The first thing done is to build a fire place of mud in the open air, which is used in common by all the pickers, as is the shed of boughs, where the food is made ready. Then, if there be plenty of wood on the place, another shed is made, which serves as a dining hall. This is made by driving two rows of stakes into the ground, the rows being about ten feet apart and the stakes about eight feet apart for a distance of fifty feet. These stakes serve as supports for a roof of boughs. The shelter is a splendid protection from the sun, but leaks somewhat in rainy weather. The barracks are used only to sleep in as a general thing, and then, as

"Three families is the rule in these shacks, one room above and one below, but sometimes four families crowd in." Outdoor dining room at side. Maryland, July, 1909. (Photo by Lewis Hine. Courtesy of Prints and Photographs Division, Library of Congress).

Mr. W. S. Crisp, one of the best known growers of the county expressed it, "they have to sleep close." On hot, dry nights the men frequently sleep outside.

The pickers begin work, as a rule, at sunrise. By this time they have finished their breakfast, and go trooping out to the berry patch to their day's task. They work under the direction of a "row boss," or overseer, who assigns them to the rows they are to pick, and keeps a general lookout to see that they don't trample the vines, pick green or decayed berries, or eat as many berries as they pick. The berries are picked in the quart boxes in which they are sold to the buyer, each picker taking several empty boxes with him as he goes in the row. When the receptacles are filled he takes them to the tally-keeper, who gives him a check for the number of filled boxes he hands in. This tally keeper is generally a bright girl selected by the grower from among the pickers. She is given a certain number of checks in the morning when picking begins, and is required to account for them in filled boxes of berries. The checks are of denominations of one, five, ten and one hundred, and change is made with them in the same way change would be made in money. For instance, if a picker brings in eight boxes of berries he is given a "ten check" and returns two "ones." Every now and then a picker's boxes are emptied into others, and if it is found that he has been picking unripe or spoiled berries he is refused his checks, and if he persists in disobeying orders he is made to leave the place.

Shanties and cooking shacks on berry farm of Bottomleys. At times, four families live in one shanty; three families is the rule—two rooms. Near Baltimore, MD, July, 1909. (Photo by Lewis Hine. Courtesy of Prints and Photographs Division, Library of Congress).

A big tin horn usually announces the dinner hour and brings about a scene of confusion. Everybody rushes to the tally keeper for checks for their berries and for some minutes she is besieged by the crowds. When the last box has been checked she makes up her accounts and goes off for her own dinner. For an hour the barracks are in commotion over the preparation and disposal of the mid day meal, after which the men sit around and smoke their pipes, while the women gossip and prepare things for supper. Then work begins again and keeps up until 5 or 6 o'clock in the afternoon. At night, when work is done, the pickers have their fun. If the farm on which they are at work is on the water the men go swimming and the little ones strip off and paddle and splash and jump around under the watchful eyes of their mothers. Then the whole crowd gathers around the barracks fire to dance and sing to the music of a fiddle or accordion, played with more or less skill. Bedtime comes between 9 and 10 o'clock, and one by one the pickers seek their corners or roll themselves up in their blankets, with the earth for their bed and the sky for a roof, and drop off to sleep, to awake at dawn and go over the same programme again. The strawberry season will be at its height in a week. Until then the workers will pick strawberries in the morning and small peas in the afternoon. After strawberries are gone they will pick raspberries, then blackberries, then big peas, then beans, and, last of all, gooseberries. The average pay is 1½ cents a quart for strawberries and blackberries, 2 cents a quart for raspberries, 20 cents a bushel for small peas, 15 cents a bushel for large peas and beans, and 30 cents a bushel for gooseberries. A good picker will average 100 quarts of berries, or $1.50 per day. The berry section of Anne Arundel is said to produce more strawberries than any other like area in the country, the average yield being about 8,000,000 quarts per season.

Eighteen years later, ***The Evening Capital*** again describes this annual migration in a special news story which appeared in the May 13, 1910 issue:

AN ARMY OF PICKERS

Annual Migration Of Strawberry And Pea Pickers To Anne Arundel Has Begun

---

POLES AND BOHEMIANS

Good Crop Of Strawberries Expected If Only Good Weather Continues.

Riding up and down the W. B. & A. electric road or the Short Line, the traveler between Annapolis and Baltimore, gazing out of the car window, sees a colony of men, women and children dotting the fields here there and everywhere.

All seem busy with their work, "What does it mean?" asks a fellow passenger. "What are they doing?" queries some one else nearby. It is the annual migration of the strawberry pickers, and soon that of the pea pickers will begin. Once a year the strawberry pickers come to the fields of

Anne Arundel county from East Baltimore. Two horse wagons in long strings are seen crossing Light Street bridge, and Haman's Bridge, loaded down with whole families and such of their wordly goods as will be needed for a sojourn of four to eight weeks in the country.

And among these worldly goods are all sorts of things. They not infrequently include a dog or two, a canary bird or a parrot, and often the youngest of the family is to be seen perched high on top of a mass of bedding clinging tenaciously to a pet kitten. Whatever is a real part of the household goes to the country when the call of the strawberry is heard and the strawberry is the trucker's first crop of the season.

Those who go to the country to pick the strawberry, pea and bean crops, as well as to do other work in the truckers' rush season, are mainly Poles and Bohemians. They come mostly from that section of Baltimore adjacent to Fells' Point. Families go together, and the farmer in getting his pickers, deals only with one (of) the leaders, who engages the others to the number required by the farmer.

One farmer requires probably twenty or thirty while another may require as many as 150 for a large farm. Quarters are provided on the farm for these families rent free and fuel is also furnished. The cooking is done mostly out of doors, and the fresh air and the fruit, equally fresh, do the children worlds of good. When the crops are gathered the pickers return from the country semi-picnic to take up work in the big canning houses of East Baltimore.

Not for a long time has there been such difficulty in getting sufficient help. Last year many people were out of work and it was easy for the farmer to get as many pickers as he needed. This year no one seems to be able to get enough. The employment agents who assemble these forces have had to cut down the numbers all around so the farmer is likely to be short handed.

Just now the strawberries in the northern section of Anne Arundel are beginning to turn red. Here and there in the fields is to be found one that is ripe all over. Two or three more days are all that are needed to make these whole fields crimson with the most luscious berry that nature affords. By Monday next the rush of Anne Arundel strawberries to the local markets will be on in full force and long, fast freight trains will be carrying the fruit to the markets at the North and Northwest.

Of the hundreds of wagon loads of strawberries taken to Baltimore from Anne Arundel, but an infinitesimal proportion reaches the Baltimore householder. Nearly all go to far-distant cities where they bring better prices and a profit to the shipper. The present crop outlook is fine, good weather being all that is needed. Young strawberry vines are full of fruit, but the old plants, those that yielded a crop last year, will produce little, having been killed off by the drought of last summer and fall.

Following strawberries will come peas, and of these Anne Arundel truckers have planted many thousand acres.

Raspberries, blackberries and beans come next, and they about complete the harvesting that is done by the caravan from East Baltimore and the Point.

The migration of the pickers is their summer outing and they enjoy it as such annually.

There were many accounts in local newspapers about the "happy, jolly" life of the pickers. But an account in the June 2, 1914 issue of ***The Evening Capital*** tells a different story:

BERRY-PICKERS' LIFE
ONE OF HARD WORK

Most Farmers Provide Quarters
Sadly Lacking in Comforts.

TYPICAL HOME OF A COLONY

Sometimes The "Row Boss" Lures
With False Promise of Big Wages.

Even though berry-pickers on an Anne Arundel county farm are forced to live amid surroundings that tend to sap their vitality, they are called upon to do the hardest kind of work.

Rising with the sun, they spend the day stooping in a back-breaking posture, ankle deep in hot sand, with the sun beating down upon them. Yet they appear to like the work, and seem to revel in the picking of strawberries and peas. They are the reliance of the farmer for gathering his berry crops. Yet in most cases they receive poor treatment at his hands.

After a day spent in his fields they return at nightfall to the shanty he has provided for them. They find a building which serves to keep off the rain and the sun, but which only affords sleeping quarters when they pack themselves in it like sardines. On one farm sixty-four pickers are employed. Including the older ones, unable to work, and the children, there are over 75 in the colony.

Inside picker's shanty, Bottomley Farm. Anne Arundel County, Maryland, 1910. (Photo by Lewis Hine. Courtesy of Prints and Photographs Division, Library of Congress).

The shanty stands in a berry field with a strip of woodland behind it. The building is a weather-boarded structure, 45x20 feet in size, and has two stories.

Within it is so crowded that one can scarcely pass along the aisle in the middle. Articles of clothing hang from all parts of the ceiling in such confusion that it is hard to see from one end to the other. The pens of bedding extend along the sides without divisions other than boards placed side up and there are no partitions or curtains. The bedding is straw on which filthy bed clothing is spread. In some cases even this is lacking. Chests in which clothing and various household articles are kept fill the center aisle.

Right: Noon hour on the berry farm in the dining room. Bottomley Farm, Anne Arundel County, Maryland, 1909.
Below: Typical cooking and eating quarters of berrypickers.
(Photos by Lewis Hine. Courtesy of Prints and Photographs Division, Library of Congress).

The sides of the walls are cluttered with all the odds and ends that cannot be left in the chests or arranged out-of-doors. A ham hangs alongside a soiled garment. A small mirror, surrounded by pictures cut from magazines or newspapers and bits of ribbon, is sandwiched in between a row of old shoes and a crucifix. At one end a narrow and steep stairway leads to the second floor. Here on one of the hottest days of the week lay an old man who had been taken sick in the field. He explained that he had a headache.

About him played five or six children, romping over the straw beds of the thirty or more persons who sleep in the place. The big door at the south end was closed to keep out the sun, thereby shutting off the air and preventing free circulation in the room. In the woods near are scattered the various family cooking ovens and tables with awnings made of branches raised over the latter. The cooking utensils are hung about on convenient saplings, while piles of wood for the cooking fires are neatly arranged here and there.

A pump over a well fortunately provides these campers with pure water. Of toilet facilities there are none.

One of the women, coming in early from the field, tells of the duplicity of a "row boss." He had brought out thirty pickers from the city under promise of big pay and better quarters. They found conditions, pay and all, the same as in former years. Four men, unwilling to be thus dumped by the "row boss," returned to the city. The rest of the party, however, being mostly women and children, were not as independent and decided to stick it out. "We were promised 25 cents an hour for extra work and we get only 10 cents," she said sorrowfully. "Extra work means weeding the rows of peas and other trucking crops."

A Polish girl and her mother, picking berries on Bottomley Farm, Rock Creek, Maryland. In the winter, they go to Dunbar, Louisiana, for oyster shucking. (Photo by Lewis Hine. Courtesy of Prints and Photographs Division, Library of Congress).

First hand accounts of many pickers paint a more favorable picture of the pickers' lives. Yet most of these accounts were from pickers who were children when they went to the farms, and no doubt through their eyes the life was seen more as an adventurous summer outing. Among the pickers on many farms were some elderly women who were too old to pick produce. They helped with the cooking, cleaning and taking care of the small children while the mothers were picking. They made life much easier for mothers, but not all farms had grandmothers to help with the daily chores. Consider the picker's routine. Up before dawn, pick in the fields for a few hours, fix breakfast, then back to the fields for the rest of the morning, fix dinner at noontime, go back into the fields, and fix supper in the evening. Clean pots, pans and dishes, wash the clothing, and care for any small children in the family. The routine varied from farm to farm and pickers got to "sleep-in" during strawberry season as strawberries weren't picked until about 9 a.m. when the dew had dried on the plants. A busy, hard day which concluded by spending the night in a crowded shanty on a bed of straw! Not exactly a "happy, jolly" life.

Apparently others felt the pickers were treated badly as the May 30, 1913 issue of ***The Evening Capital*** ran the following story previously printed in ***The Baltimore News***:

ATTACKS MEN WHO
HIRE BERRY-PICKERS

Rev. Joseph Wise Declares Hands Are Forced To Live Like Cattle.

CALLS GOV. TO INVESTIGATE

Preacher Says City Suffers From Condition Which Exists In County At Present.

Farmers of Anne Arundel County who send money for foreign missions to help the heathen would do well to use the money at home to give the berry-pickers human conditions of habitation," said Rev. Joseph Wise, a Methodist Polish missionary, who has just made a visit to many of the patches and has inspected the shacks.

This year and last year Rev. Mr. Wise has visited in all over sixty shacks and has found almost all of them worse than any laboring conditions he has met with in any foreign country. He believes that in some respects the conditions are worse than those in the heathen lands.

"What I am saying now," he declared," I have said to the faces of the farmers themselves. I have preached in churches there and have condemned from the pulpit the conditions, but even the best of the farmers seem to think that anything that is good enough for a dog is good enough for Polish berry-pickers. The Governor of the State should appoint a commission to investigate conditions in Anne Arundel shacks, if he wants to know what is one of the chief causes of increasing immorality and vice.

"Men, women and children, without regard to modesty, are compelled to live together and to undress and dress together. The people by the hundred are herded just as the cattle are herded, and

if anything the cattle are the better cared for. The poorest of the people who go to pick the berries have a sense of shame and a love for decency. This they are compelled to lay aside. In the foreign countries from which they come there would not be permitted any such conditions.

"The people are reduced to the level of animals. It means great distress to these people. They feel the shame of it keenly. Out of it grows the worst forms of vice. The farmers seem to throw the people together without regard to consequences. The farmers know these consequences. They learn of the results, and they say only: "Well, the people are just like dogs." They have told me that.

"Here is one reason the people of the city should try to enforce through State legislation some higher humanity among these employers. The vice that starts on these farms counts against the city eight or nine months a year. People who are degraded on these farms and in these shacks are that much harder to be dealt with back in their homes. The good that missionaries and other agencies do at other seasons is undone at this time. The people of Baltimore through official and unofficial agencies, suffer and pay the price for this wholesale damage of character.

"If anything is done it must be done by the Governor of the State or by the Legislature through an investigation. The farmers, themselves, although devoutly religious and generous in supporting foreign missions, refuse to listen to any suggestion for improving conditions. Knowing as I do what the conditions are, I try to prevent the people going to those places to work, and other social workers join in trying to keep the people in the city.

"But the poor people need this money and they earn it. They deserve to have better conditions and the State should rise up and protest against the present outrages that are committed on these farms."

The farmers, no doubt, were upset by Rev. Wise's allocations, but were not willing to make any changes. Pickers had been coming to the Anne Arundel farms for over 40 years and were seemingly content with their primitive living conditions. From most accounts, written and oral, the pickers were a "happy, jolly" lot and were willing to make sacrifices to earn their wages. Many pickers returned to the same farm year after year and knew full well what they were getting into when they boarded the farmer's wagon heading to the country. Many pickers bonded with the farmers. Long after the truck farms were gone, some pickers returned to the farms to visit with the farmer and his family. Quite a few pickers eventually bought their own farms in Anne Arundel and recruited pickers of their own.

Occasionally disputes would surface between the farmers and the pickers. One such dispute occurred in June, 1914, when pickers demanded 20 cents a bushel for picking beans on the farm of James S. Smith. Mr. Smith refused the demand as pickers were getting 15 cents a bushel from all the farmers in his vicinity. The pickers went on strike and refused to do any picking. Mr. Smith obtained a warrant compelling the pickers to vacate his farm in three days. The pickers refused to leave and Mr. Smith notified the police. By the time the police arrived, half the pickers had left the farm. At the request of Mr. Smith, the police were detailed to guard the premises over night. The farmers finally agreed to pay the 20 cents for a bushel of beans as otherwise they had no one to pick the beans.[4]

During World War I, the farmers were again confronted with an acute labor shortage. From ***The Evening Capital***, May 16, 1917 we find:

THE SCARCITY OF BERRY PICKERS WORRIES FARMERS

Weather, War And High Cost of Living Are Blamed For The Resulting Situation.

The unusual weather, the high cost of living and the scarcity of labor occasioned by the war have had their effect upon the strawberry and vegetable farmers of Anne Arundel county. Usually about the middle of May there is an exodus of members of the Polish colony in East Baltimore to the farms of Anne Arundel county for the strawberry picking season, and agents of the farmers have been at work in trying to get berrypickers for the last week. They have met with very little success. Many of the men and even some of the women who formerly would work on the berry fields of Anne Arundel are now employed in more lucrative positions. Quite a number are working in the munitions factories, while others are at other occupations which bring to them greater profit than they would reap on the berry farms.

Usually these berrypickers would carry with them the food to last them to the end of the season, but now, it is claimed, they are not able to purchase sufficient food at the high prices charged.

In addition to the scarcity of labor, the pickers the farmers could recruit demanded to be paid two cents for a quart of strawberries rather than the usual 1½ cents per quart. About 300 farmers of Anne Arundel County met in a war conclave at Haron Station and agreed to meet the demands of the pickers.[5]

In May of 1917, to rally farmers behind the war effort, Governor Emerson C. Harrington of Maryland held a patriotic meeting in the interest of food preparedness and cultivation of the farm under the auspice of the Men's and Woman's Survey and Preparedness Commission of Anne Arundel County. Held in the House of Delegates in Annapolis, the meeting was attended by a large crowd as many extra chairs were brought in, and the galleries were filled.

Mrs. Theodore Johnson, chairman of the Anne Arundel Woman's Preparedness and Survey Commission, announced that she has placed a large order for canned goods and encouraged women to plant and can farm products as they have never done before. "America expects not only all men, but every woman to do her duty."

The Governor made a magnificent speech outlining Maryland's rich history and urged women to increase their efforts in home food-gardening. He then addressed the farmers: "Mr. Farmer, you are helping your country as patriotically as those who are marching. Mr. Laborer, you stay upon the farm and work there."

The Governor emphasized that this is a Farmers' War, and that Anne Arundel County has much waste land and needs to produce more food. The farmer at last had come into the limelight.

Professor Thomas W. Symons, of the Maryland State College of Agriculture, called upon the farmers to rally and told them that the success of the conflict will depend upon the farmer. He reported that in Anne Arundel County there were 2,038 farms containing 208,000 acres of land, but 69,000 acres were not under cultivation. He pleaded that they needed to increase their productivity and put the waste land under cultivation.

Professor S. C. Richardson, of the Maryland State College of Agriculture, also talked to the group, telling them that there was a shortage of food in our country and it was up to the farmers to prevent its soldiers from fighting on short rations.

"This is a Farmers' War and it is going to be won by men in the furrows and not by the men in the trenches. If you put into your work on the farm patriotism, the victory will be won," and he concluded with, "Let every farmer nail that magnificent old flag to his plow and to the tune of 'Tramp, Tramp, Tramp the boys are marching, 'sing' Plant, Plant, Plant the boys are marching.'"[6]

Four days later ***The Evening Capital*** carried an article headed, "**Can, Can, Can, The Girls Are Canning,** which urged each woman to start a miniature canning factory in their own kitchen..."[7]

The farmer had never enjoyed such attention. But labor was critical, and the buyers of truck produce refused to pay the farmers the premium rates demanded by an inflated war economy. Despite a critical need for more food, the farmers of Anne Arundel County were forced to destroy some of their crops as the purchase price from produce buyers and packers was less than the farmer's production price.[8]

Anne Arundel truckers were receiving as low as 60 cents a bushel for peas, while the costs for the farmers to produce a bushel of peas was one dollar. Strawberries were selling for 4 cents a quart while the farmer's production costs were 5½ cents.[9]

Oregon R. Benson, Jr., representing the farmers of Anne Arundel, said that the farmers would rather give their produce away than take it to the market at the loss they are now sustaining. He stated that the first crop had already been plowed up which has created hard feelings toward the farmer.

Mr. Benson said, "We do not want to fight the packers. All we want is a fair price and we have to get it or quit."

"Those boys of ours across the sea must not know that good food is being plowed under because we cannot get enough money for hauling it to market. The Government should fix the prices."

J. W. Disney, Jr. of Severn, Anne Arundel County, who had been farming for 40 years, said that he had been running his place on his bank account and there was nothing going back into the bank.

"The laws of supply and demand seem to have become reversed," Mr. Disney added. "Never before has food been so scarce, and acting on the Government's plea we have all planted more and worked harder than ever. Yet the prices are the lowest in 15 years. Labor has gone up 200 per cent in the last two years; phosphate has risen 100 per cent in the last year, and it costs us 30 cents now to pick a bushel, as against 20 cents last year."

To make matters worse, the government placed rather stringent regulations on enemy aliens. The leader of the Polish colony in East Baltimore said that in previous years from 3,500 to 4,000 families would go from East Baltimore alone to the county districts, where they would remain practically the entire spring and summer seasons, starting with strawberries and ending with the picking and canning of peaches and tomatoes. Because of the scarcity of labor and of government regulations, he felt that not more than 500 families would be available.

"It is an injustice," said Mr. Disney, " for the government to class the Poles as enemy aliens. We have our fighting legions now in France, and we are encouraging in every way possible enlistments of Polish young men in our army here. The Poles who are fighting the cause of Germany and Austria in this war are not fighting from choice, but because they are compelled to. We have no more love for Germany and Austria than we have for Russia. All of these powers have exploited the Poles..."

He stated further that he would like to see Congress or the President lift the ban against Poles of Austrian or German origin. No more patriotic class of foreign-born exist in this country today, he said, and if this ban were lifted it would go far to solve the labor problem on our farms this summer, particularly in relation to berry picking and kindred pursuits which relate to the canning industry.[10]

To make up the labor shortage, Governor Harrington issued a proclamation calling youth between 16 and 21 years old to the colors for food production during the summer.

Frank B. Cahn, director of the Boy's Working Reserve for Maryland, made rounds at various high schools calling for volunteers. Maryland hoped to recruit 2,500 boys to work for not less than six weeks on a farm. Volunteers must have their parents sign their registration card and must pass a physical examination.

After a boy enlists, he is given a button, showing his enrollment. After working six weeks, he is given a bronze badge, showing his service to the country. After ten weeks service, he is given a United States chevron. Pay for inexperienced boys is $15 for the first month and $30 a month after that. Those experienced get $30 a month from the start. All boys will receive free room and board.[11]

Other programs were initiated to help the farmer with his labor problem.

The military draft exempted skilled farm laborers and gave furloughs to soldiers who had farm experience so they could work on the farms for a short period.

Men with farm experience living and working in towns were urged to put aside their city labors for a few days and work in the farms, especially during critical periods in planting, cultivating, or harvesting. This was not considered the best source for farm labor as the town men were not "hard," but they could furnish the farmer with the necessary manpower for emergencies.

Women were recruited to work on the farms in many of the "lighter labors," such as vegetable cultivation and picking truck produce and fruits. The picking force in Anne Arundel County was always made up largely of women who were in many cases more efficient than men. Women were also called upon to relieve the men in the towns so that they could go to the farms for a few days.

Another source of labor targeted by the government was from the large class of idlers and loafers. Local sheriffs and town marshals were asked to rigidly enforce local laws against vagrancy and loafing. Every man, whatever his status or condition, was needed in this time of emergency to put in full time.

The government also waged patriotic campaigns urging its citizens to support the war cause and work more efficiently to help bring victory for the allies. The government cannot create labor, or conscript labor to the farms, but they can stress the importance for communities to unite and solve their own problems.

Despite these efforts, the farmers never did obtain enough labor; nevertheless, through hard work and long hours, the Anne Arundel truck farmer rallied to the needs of the country and increased his production to help the war effort. The packers and commission agents raised their prices and, with the special programs to help the farmer with his labor problems, the farmers produced more with less labor.[12]

From ***The Evening Capital***, May 28, 1918:

ANNE ARUNDEL BERRY
SEASON HAS ARRIVED

Women And Children Are Doing Most Of The Picking Of The Fruit This Season.

The aroma of strawberries from the farms of Anne Arundel is to be met all about town these days. On the stalls in the market, about the fronts of stores and from the wagons of hucksters, there is to be noticed this delicious odor.

Berry pickers, speeding away in auto trucks, are seen daily. The trucks are loaded with household furniture, women and children, but there is a notable absence of men as in previous seasons. The latter find more lucrative employment nearer home.

From ***The Evening Capital***, June 5, 1918:

HOW FARMERS CAN GET SOLDIERS HELP

Must Apply To Draft Boards And Name Men They Want On Farms.

If the farmers of Maryland want soldiers to help them on their farms this summer, each farmer...must file his application with his local draft board, which will forward the application to Camp Meade.

The farmers must name the men they want. Knowing these boys who went from each district, this provision makes it easy for the farmers to get men they know. The soldiers then make application for furloughs under the general order of the War Department, which permits them to engage in agricultural work.

From ***The Evening Capital***, June 14, 1918:

BOY FARM VOLUNTEERS
MAKE GOOD AT WORK

Letters Declare Youth "Are Quick To Learn And Are Doing Well In Counties."

The farmers of Maryland are more than pleased with the work of the boys who have been assigned to them from the ranks of the Boys' Working Reserve, according to Frank E. Cahn, director of the reserve. Every day letters come in from the farms to which the boys have been sent, commenting upon their work, their eagerness to learn and their skill at picking up and applying knowledge which to many of them is new.

Most of these letters are followed up by others from farmers in other parts of the State, asking for boys to help harvest the crops. So far, 227 boys have enlisted, and Mr. Cahn estimates that he can use about 600 more.

This was well short of the 2,500 boys they had hoped to recruit, but the effort helped the farmer as every set of hands contributed to the harvesting of more produce.

From oral traditions, one would be led to believe that all Anne Arundel pickers were of Polish descent. Yet frequent mention is found in contemporary newspapers of Bohemians and Germans working in the fields. In reality, there were many European countries represented on Anne Arundel farms; yet time has tended to blend all European pickers into Poles. The immigrants from Little Poland probably dominated the picker force, and on some farms Poles were used exclusively.

The contribution made by African-Americans to the success of Anne Arundel truck farming was very significant, yet often overlooked. Many farmers employed African-Americans on a year-round basis and used them primarily to perform the infinite number of daily chores required for any farm to operate, in addition to picking produce. There were also a number of African-Americans who owned their own farms and were dedicated to truck farming.

However, many African-Americans were used in the capacity of migrant pickers as a large number left the cities and migrated to the farms as the sweet aroma of the strawberry filled the air. The following article appeared in ***The Evening Capital*** in its first year of publication on May 24, 1884:

STRAWBERRY PICKING

At best it is a difficult thing to procure a good servant, but now when picking has begun, the culinary department in every other household in this city is deserted, and the lady of the house is compelled to take upon herself the onerous duty of cooking during these hot days. The cause is every man, woman and child, of the colored population, is seized with the strawberry picking fever, and most of them desiring a change of scene, have gone into the country for this purpose. - Consequently the ladies will be servant hunting for months to come. - For there is a complication of diseases which arises from this picking fever, for after the strawberry, come the raspberries, whortleberries, blackberries, gooseberries which affords a summer of work for the pickers who, like their mistresses, prefer to leave the city for the country on the approach of Summer.

African-Americans were imported not only from Annapolis and Baltimore, but also from the Eastern Shore, as described in an article from ***The Evening Capital*** of May 19, 1900:

STRAWBERRY PICKERS MECCA

A sloop and a bugeye left Cambridge on May 17 carrying 100 colored people to Anne Arundel County to engage in strawberry picking. They are the first to leave and it is expected that several hundred more will follow in a few days.

In 1900, African-Americans were imported from the Eastern Shore, but by 1916, they were recruited from Anne Arundel County for the Eastern Shore. From ***The Evening Capital***, May 26, 1916:

THE STEAMER STARLIGHT SAILS WITH BERRY PICKERS

The steamer Starlight sailed on Wednesday afternoon from the Tolchester Company's wharf, foot of Prince George street, with a party of colored berry pickers.

They were secured here and in the suburbs through the agency of Mr. Richardson, for picking berries on the Eastern Shore. It was desired to secure 500 berry pickers from Anne Arundel, 250 of whom were wanted for Denton, Caroline county and 250 for Salisbury, Wicomico county. Mr. Richardson was not able to secure but 120, and these left here on the Starlight Wednesday. They are promised good wages and steady work for some weeks.

A number of colored folk who went on the berry picking expedition gave up jobs here to engage in the work on the Eastern Shore.

With the hot humid Anne Arundel summers, it is easy to understand why one would look forward to leaving the city of Annapolis for the cooler country farms. From these articles we can assume that a substantial number of African-Americans worked on the farms during the picking season. As truck farming increased on the Eastern Shore and decreased in Anne Arundel County, more and more workers migrated to the Shore during the picking season.

In accordance with the customs of the times, pickers were segregated and rarely would the African-Americans and Baltimore migrants be found picking in the same fields or living in the same shanty. Most farmers would not even allow their own families to have contact with the pickers regardless of their race or nationality. Nevertheless, a strawberry picked by an African-American was just as sweet as a strawberry picked by an immigrant.

STORIES AND LEGENDS

The picker's life was surrounded by numerous stories and legends, some of which have been told by the people who worked the farms, farmers and pickers.

From Mrs. Maynard - whose father farmed on Burns Crossing Road: "Pickers stayed in our old barn. Each family had a stall. One Polish woman came to the farm very obviously pregnant, but she didn't miss a day of work. One afternoon the rowboss noticed she was no longer pregnant. She had given birth to a baby girl on her dinner break and was back in the field in the afternoon. They tried to get her to take a few days off, but she wouldn't hear of it...and kept on picking."

From Margaret Rose - granddaughter of James P. Watts and daughter of Penrhyn S. Watts, truck farmers: "People from surrounding neighborhoods picked our produce and my father used picker checks. One picker was totally blind and walked with a white cane. He picked string beans and did a good job. He wouldn't pick strawberries because he said he may mash them."

From Jefferson Monroe Chairs - In the 1930's, Mr. Chairs served as a rowboss at the age of 15. This was an era when the traditional immigrant picker system had almost disappeared. Farmers used mostly local labor for pickers and rowbosses: "I worked as a rowboss for George A. Jenkins at the age of 15. I supervised pickers and handed out checks as they turned in their produce. I wore a carpenter's apron and kept a variety of checks in the pockets. I remember handing out the heavy brass tokens first to lighten the load. The aluminum tokens were given out last as they were much lighter. Most of our pickers were blacks and were transported to the head of Rock Creek by steamboat from Baltimore."

From Anna Browskie - rowboss for Harry Owens, as quoted from the ***Maryland Magazine***, Summer 1978, and from personal interviews: "My husband Peter and I were rowbosses for Harry Owens for 20 years. After Peter died, I served as rowboss for two more years. We were paid 25 cents per picker. The rowboss was paid $30 per week. If any pickers went back to Baltimore, we had to give back a portion of our salary. The shanty was one long building which slept 24 families, 12 on each side."

"One night the older women complained of the noise of the teenagers singing and laughing late into the evening, under the apple trees. I put on a sheet, draped it over my head and went out to scare the kids. When the kids saw me, they screamed and started throwing stones at what they thought was a ghost. I ran off and decided to go to bed. Later, the kids came in talking about the ghost they had scared off. They went to sleep, but I was awake half the night rubbing my bruises."

"The rowboss would let us know, as early as Christmas, when the farmers would need their pickers. On the designated day, we were ready with our clothes all packed, food which required no refrigeration and a featherbed. We knew we were going to the farms to work, but it felt like we were going on a vacation to the country. About thirty of us would pile onto a wagon and we would laugh and sing as we headed for the country. One year it was raining and only a few had umbrellas. Everyone got wet, but it didn't seem to matter."

"In the evenings, after supper, we made music any way we could. We had accordions and fiddles, but we also played combs and clapped hands. My son Stanley always brought his saxophone, and we danced the polka, holding hands, making circles. Young boys played war in the woods with sling-shots and lovers lingered in the shadows. Those evenings were a happy time."

Stanley Browskie - son of Anna Browskie, remembered the sling-shot wars that the kids had in the woods and that they dammed a stream and made their own private swimming hole. He also recalled that with all the teenage boys and girls, there was lots of smooching going on. [13]

From Ora Smith regarding Hamilton Smith's rowboss, Mr. Scheaffer:

"One day the market commissioner called Ham Smith to tell him that the beans had dropped to rock bottom price, so they decided to tell the pickers to stop picking. The rowboss went into the field and told the pickers to stop picking. The pickers were so enraged that they jumped on the rowboss and beat him up."

From ***The Evening Capital***, June 3, 1895:

Joseph Lavicki, a strawberry picker on the farm of Mr. John T. Hammond, a short distance south of Brooklyn, Anne Arundel county, dropped dead while at work Saturday. He was fifty-eight years old and lived on Fountain Street, Baltimore.

From Isabel Shipley Cunningham - granddaughter of Larkin Rodolphus Shipley and daughter of Marvin R. Shipley, truck farmers: "On Saturday afternoons, when the week's baking was completed, the wife of the rowboss came to our house with crusty loaves of bread, still warm from the brick oven. When she left, she carried big bunches of peonies, iris, or roses which brought forth voluble thanks in broken English, as the pickers loved flowers as much as music. Sundays were happy times because the husbands and fathers who were employed in Baltimore dressed in their best clothes and came by train to visit their families. Young couples took walks along the country lanes, picking bunches of daisies, Queen Anne's lace, and other wild flowers, and always there was music."

Larkin Rodolphus "Dol" Shipley and wife Isabel. (Courtesy of Isabel Shipley Cunningham).

Lewis Hine Photos

Lewis Hine, a photographer, took pictures of children in labor environments for the National Child Labor Committee in 1909 and 1910. One series of photos was taken on the Bottomley truck farm on Rock Creek in Anne Arundel County. These are the only known photographs of the picker's life on an Anne Arundel truck farm. The captions on each photo were written by Mr. Hine. The series of photos are from the Prints and Photographs Division of the Library of Congress. These photos document that some of the pickers were migrant workers as Mr. Hine states that several of his photos are of families who migrated to Mississippi or Louisiana during the winter. From tradition, most of the Anne Arundel pickers were not migrant workers as they had permanent homes in Baltimore and moved to the farms only during the picking season. Mr. Hine's photos are also the only known photos showing both the inside and outside of a picker's shanty.

Right: Mrs. Lissie and family (Polish). They all work in the fields near Baltimore in summer and have worked in Biloxi, Mississippi for two years. Bottomley Farm, Anne Arundel County, Maryland, 1909.
Below: Slebzak family (Polish) working on Bottomley farm. They have worked here 3 years and one winter at Avery Island, Louisiana. All work except the very smallest. She hangs around the fields. Work begins about 4 a.m. and work until sunset. (Photos by Lewis Hine. Courtesy of Prints and Photographs Division, Library of Congress).

Child Labor Study

In 1923, the U.S. Department of Labor, Children's Bureau, issued a study entitled, "Child Labor on Maryland Truck Farms," by Alice Channing (Bureau Publication No. 123). The study was the result of a survey Ms. Channing had taken on truck farms in 1921. The purpose of the survey was to study the work of children on truck farms and how this work affected their schooling. A major portion of the study deals with pickers in Anne Arundel County and gives vivid descriptions of their life on the Anne Arundel farm.

The study includes both resident children and migratory children working on the farms. Over 800 children were interviewed. One-third were children of farm owners and tenants, one-third were children who were hired as day laborers and who lived near the farm, and the remaining one-third were children in migratory families performing seasonal work. Of the 540 resident children, 322 were black, most of whom were children from families with fathers working as laborers on farms, the roads, the Baltimore docks, or at nearby military camps. About one-third of the black children came from nearby smaller farms where parents would allow their children to work on larger neighboring farms. It is also interesting to note that about one-half of the white farm owners were foreign born, chiefly Polish and German.

Laura Petty, a six year-old berry picker. (Photo by Lewis Hine. Courtesy of Prints and Photographs Division, Library of Congress).

Picking strawberries, peas or beans requires little skill and can be done by children as soon as they are old enough to distinguish a mature pod or a ripe berry from a green one. Consequently, many children were used on the truck farms to pick the produce. The study included information from 145 migratory families, two-thirds of whom had worked on the farms three or more seasons and 21 families had worked on the Anne Arundel farms for eleven or more years. After the harvest, the workers usually returned to the city and only a few families followed the work from the berry fields of Anne Arundel County to tomato or other kinds of canneries. Some went as far south as Alabama, Louisiana, or Mississippi.

Most of the migratory workers were Poles and the fathers of 126 of the 145 families had been born in Austria, Germany or Russian Poland. They were not recent immigrants as only one father and four mothers had been in this country less than ten years; ninety per cent of both the parents had lived in the United States 15 or more years. Living in the segregated Polish districts in Baltimore, the mothers in 85 families and the fathers in 43 families had not learned to speak English. A large proportion of the immigrants were illiterate.

Due to the large unemployment of the early 1920s, 81 fathers were with their families on the farms, which was unusual, as most fathers stayed in the city where they held full-time jobs. The families did not rely on their farm earnings to carry them through the year. They would usually carry back to the city a family earning of between $100 to $200, which was used to buy supplies for the winter: flour, potatoes, coal and sometimes clothing.

Twenty five migrant camps were visited for the survey. In these camps were 268 families, consisting of 1,074 persons, including approximately 550 children under 16 years of age. Most of the migrant workers lived in a two story shanty, with one room on each floor. The floor was divided into sections or pens by boards about ten inches in height. Each pen was about six feet long and four to six feet wide and covered with straw for a mattress. Each family was allotted one of the pens and at night men, women, and children, partly clad, lay side by side, separated from the next family by the ten inch plank.

Wooden chests located at the foot of the pens and a shelf around the room used by the pickers for miscellaneous possessions were the only furnishings. Clothes were hung on nails driven into the walls. In some shanties attempts for privacy were made by hanging sheets or blankets around the pen. The shanties often had open windows with no glass or screening and the families would sometimes hang cloth over the windows to keep bugs and flies out. This also shut out any fresh air which might come through the window. In some camps chickens were left to scratch through the straw of the beds. The shanties would house as many as 100 pickers of both sexes and all ages.

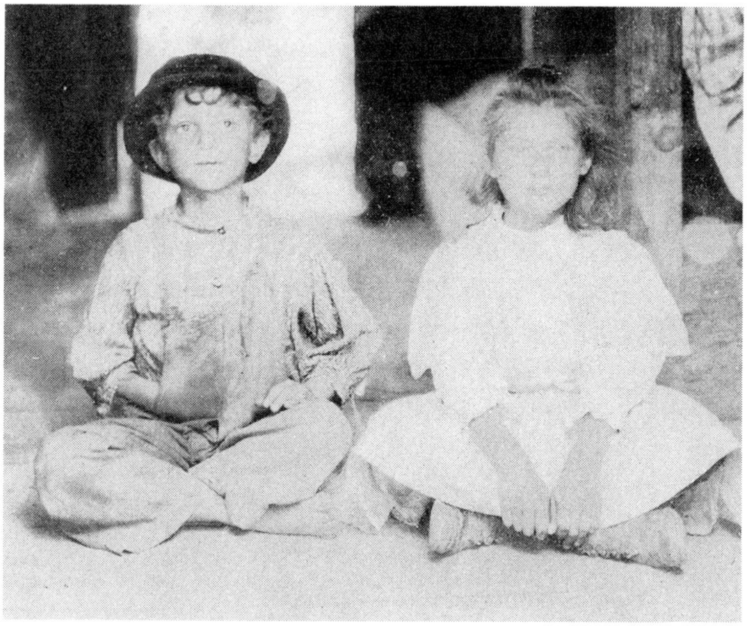

Above: Berry pickers at Bottomley Farm, 1909. Left: These children are representative of the families that occupy this one room shack on Bottomley's Farm. There are only sliding curtains to separate families. (Photos by Lewis Hine. Courtesy of Prints and Photographs Divsion, Library of Congress).

Cooking and eating arrangements were fairly adequate. The farmers provided fuel for cooking, lumber for making tables and benches, and bricks and sheet iron for stoves. Each family had its own stove and a table set up in the shade of the trees surrounding the shanty. The pickers made community ovens and their delicious homemade bread was sometimes shared with the farmers.

Very little attention was given to sanitation. Waste water was thrown where it was used and most of the garbage was thrown in a nearby ravine or in a stream, if convenient. The garbage areas were used from season to season and were not cleaned between seasons. Old tin cans and other refuse often made the area very unsightly, and swarms of flies feasted on garbage.

More than half the families had no toilet facilities. The odors pervading the camps were offensive. Of the 25 camps visited, only one had toilet arrangements considered adequate with three well-constructed privies which were cleaned twice a week by children assigned by the rowboss. The privies were often constructed by the pickers by placing several boards together over a brook or a stream and shielded by a piece of burlap stretched across the branches of a tree. There were no pits and the privies were usually within 50 yards of the shanty. Some were dangerously near the water supply and often the stream used for the privy was also used for washing clothes. The privies were foul smelling and infested with flies, and pickers frequently used the woods in preference to using them. Where there were no privies, the pickers would use the woods.

Water sources varied. Some camps had a spring or brook for both drinking and washing; some had dug wells and only a few had access to a drilled well. In one camp the water supply was contaminated and the farmer hauled two barrels of water a day from his well for 30 pickers. When the women complained that they did not have enough water for washing, the farmer told them to "wait until it rains."

Four year old boy in berry field of Bottomley Farm. (Photo by Lewis Hine. Courtesy of Prints and Photographs Division, Library of Congress).

African-American children from tenant farm family, Shipley's Farm. (Courtesy of Jack Kelbaugh).

None of the pickers appeared to fear polluted drinking water, but criticism was often voiced over the lack of toilets, overcrowding, and the lack of privacy. One rowboss expressed himself in broken English, "Women, boys, and girls sleep all together. No good." Another family said that their children learned bad habits from the others and that, although lack of employment had caused them to come to the farm this year, they would not try it again. Many families described the way in which they lived as "like hogs," "like sheep," and "like cattle beasts."

Yet camp life was considered by some workers as a vacation, and the fresh air was said to be "good for the children." They played games, went wading in the brooks or swimming in the creeks, used the swings which were sometimes provided by the farmer or the hammocks brought by the pickers from Baltimore. Mouth organs, accordions, violins, and banjos were often heard in the evening. One camp had a grafanola. Dancing and singing was popular in the evenings. One father who had taken a furlough from his steady job on the railroad said that being in a camp "was the nearest to a picnic that he had."

On Sundays many of the fathers who stayed in the city came to the farm, sometimes bringing food, newspapers or candy for the children. Only a few families went to church and in some camps a few families held their own church service in the woods. Normal working hours were from 4:30 to 8:30 a.m.; 9:30 a.m. to 12 noon; and 3 p.m. to 6 p.m., but these hours varied from season to season depending on the state of the crops. For strawberries the starting hours were later as the dew had to evaporate before the berries could be picked. During seasons with poor crops, as 1921 was, the pickers were discouraged as their pay was considerably less than in a good year.

The usual rates paid to pickers in 1921 were for strawberries, 2 to 2½ cents per quart, for beans, 25 to 30 cents a bushel, and for peas, 30 to 35 cents a bushel. For weeding, children were paid 10 cents an hour.

When any picker emptied their bushel of beans, he or she was given a small brass check. The children's checks were usually held by the parent, and at the end of the week, or, more likely, at the end of the season, the checks were redeemed by the farmer. From its daily labor, a typical family of four would make from $3 to $6 depending upon the condition of the crops.

The study found that the children working on the farms were illegally losing 30 or more school days which was seriously hampering their education. Black children were missing less school because their school year had fewer required days and black schools were closed during May and June (black schools were in session 138 days, as compared to 187 days for white schools). The study called for some public supervision in order that growing children might escape the physical and social effects of promiscuous and unhealthful living conditions.[14]

CHAPTER FOUR NOTES

1. Information on Civil War letter courtesy Jack Mellin of Sherwood Forest, MD.
2. Fairbanks, W.L., *A Statistical Analysis of the Population of Maryland,* Maryland Development Bureau of the Maryland Association of Commerce, Baltimore, 1831.
3. *The Sun,* May 17, 1918, "Truckers Are Puzzled." Estimates 3,500 to 4,000 families go from East Baltimore to the country to pick fruits and vegetables.
4. *The Evening Capital,* June 16, 1914, "Bean Pickers on Strike."
5. *The Evening Capital,* May 18, 1917, "Make Concessions to Army of Berry Pickers."
6. *The Evening Capital,* May 23, 1917, "Drastic Food Bill Now Before House."
7. *The Evening Capital,* May 27, 1917, "CAN, CAN, CAN, The Girls Are Canning."
8. *The Evening Capital,* June 11, 1917, "Farmers Anxious For Food Control."
9. Ibid
10. Ibid
11. *The Sun,* May 14, 1918, "Called To Land Colors."
12. *The Evening Capital,* May 25, 1918, "Some Suggestions For Farm Labor."
13. Gilbert Sandler, *Maryland Magazine,* Summer 1978, "Pickers Checks, Currency of the Crops."
14. Alice Channing, *"Child Labor on Maryland Truck Farms."* U.S. Department of Labor, Children's Bureau; Bureau Publication No. 123, Washington, Government Printing Office, 1923. This study concentrates on truck farms in Anne Arundel County. It details conditions found on the farms specifically dealing with resident, migratory and African-American children.

The Checks

CHAPTER FIVE: THE CHECKS

Picker checks are very plain, bearing only the impressions of the steel die, hand-stamped initials of the farmers and a value. They were made for utility and not for promotion or advertising as is the case of many tokens. Numismatically, they are not exciting for if you have seen one series, with few exceptions, you have seen them all. Yet plain as they are, the predominantly brass tokens offer enough variety in size, shapes, border design, materials, initials, and values to be of more than casual interest to the token collector (exonumist). Each token is unique, as they were hand-punched with initials and values and each token was handled by the farmer, the rowboss and the picker, whether they were Polish, Bohemian, or African-American. They have a strong sense of history that defines a culture lost in our time, but which helped define who we are today. They also reflect a pride in the farms, the farmers, and the pickers.

Here is a series of tokens which is pure Maryland, and they are the largest single series of Maryland tokens in existence. There are well over 1500 known varieties of picker checks from over 400 different farmers, mostly from the northern half of Anne Arundel County, Maryland. But contained within these tokens is the story of the vanquished truck farms, the hard life of the farmers and the hopes and dreams of immigrants. In many cases the tokens are all that remain of a culture which dominated Anne Arundel County for almost a hundred years.

The checks were given to the pickers as they turned in the produce they had picked. At any time, a picker could turn in the small value tokens for larger values to lighten the load. The checks could be used as money in the farmer's commissary if he had one or spent as money in many local stores. The Parker's Third District store accepted the checks from pickers in lieu of money. The Parkers would contact the farmer to redeem his checks, or by arrangement the farmers would stop by the district store at the end of the season and redeem his checks with cash.

Carroll M. Cole's father, Wilbur Cole, farmed in the area now occupied by BWI Airport. In the early 1930's he was using Polish pickers, but had stopped using picker checks. All the old checks were stored in the attic. A local merchant called and asked Mr. Cole to come and redeem his checks, but he said he wasn't using checks anymore. The merchant said, "Well I got some!"

It seems that a young Polish lad was sneaking some of the brass tokens out of the attic and taking them to the store to spend. They caught the youngster who turned out to be the son of their rowboss.

Wilbur Cole put an end to the foolishness by taking his checks to Baltimore and selling them for scrap, which is why "RSC" checks used by Richard Stanley Cole, Wilbur's father, are not found in any collection of picker checks.

Information concerning the tokens is scarce. Not all have been rediscovered, as more undoubtedly still lie hidden in attics, basements, barns, or lie buried under ground. Some will never be found as they were sold for scrap metal or thrown into old wells or privies and are lost forever. Many relatives of the farmers still cling to their tokens, and rightly so, for these checks represent their truck farming heritage which they wish to pass on to their families. Some of these families have shared their token information, but some will not. More tokens will certainly come to light in the future.

In Maryland, pickers' checks were used predominately, but certainly not exclusively, in Anne Arundel County. Checks were used in several other counties surrounding the city of Baltimore and on the Eastern Shore, but most of the metal Eastern Shore checks were used in canneries or packing houses. Most farmers on the Eastern Shore used small printed cardboard chits which are found in different colors corresponding with their values. A few brass pickers' checks are known to have been used in New Jersey, New York, and Virginia, but not nearly to the extent as used in Anne Arundel County. In all probability similar checks were used in most of the large truck farming areas, but never has as great a concentration of checks been found as in Anne Arundel County; hence the name, Anne Arundel County Pickers' Checks.

S.W. Fletcher, professor of horticulture at the Pennsylvania State College, in his book, *Strawberry-Growing,* written in 1917, describes the methods farmers used to keep their accounts with their pickers. He mentions cash at picking time, day-book records, redeemable checks and redeemable punch tickets. "Small cards or checks have been used more than any other method of accounting but are now superseded by punch tickets, except when the number of pickers is less than twenty-five. Checks are pieces of cardboard about one inch by two inches, with the name of the grower printed on one side, together with the number of quarts it represents...Sometimes aluminum checks, called "strawberry money," are used."[1] Professor Fletcher's casual mention of aluminum checks and omission of brass checks, suggests that metal checks were not commonly used by strawberry growers. This supports the supposition that the high concentration of brass checks used by the farmers in Anne Arundel County and the Baltimore area is unique to this region.

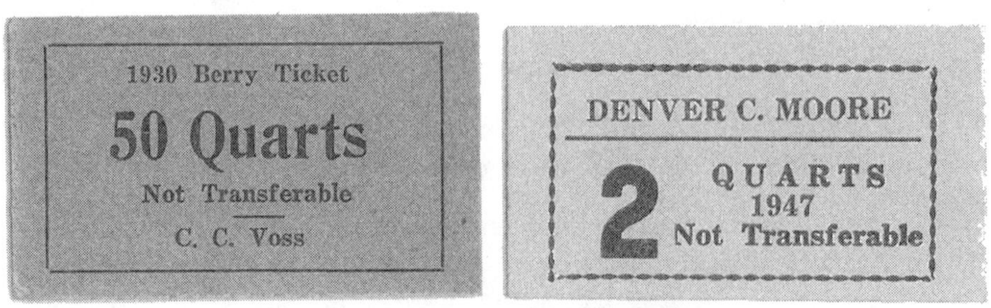

Eastern shore cardboard checks

The listings in this volume are as complete as is now possible. There are few serious collectors of picker checks and most of their holdings have been inventoried and are included in this catalog.

Anne Arundel pickers' checks came into prominence through the efforts of Ora Pumphrey Smith. Working as a volunteer for The Ann Arrundell County Historical Society, Ora Smith and her husband, John, made a few of the Smith family's checks into jewelry to sell to raise funds for the society. They polished the brass tokens, drilled small holes in them and strung them onto a gold chain to make a bracelet. The demand was so overwhelming that Mrs. Smith issued a call for members and friends to dig into their attics, basements, or old barns to find more tokens and donate them to the society to make more jewelry.

Ora and John Smith making bracelets out of picker checks. (Courtesy of Ora and John Smith).

Tokens surfaced from all over the county. They received as many as 2000 tokens from a single farmer. Some were recovered from old wells and garbage dumps. The Smiths expanded their jewelry creations, making earrings, necklaces, tie tacks, and scarf holders and, over the next twenty years, made over $60,000 for the society.

Mrs. Smith kept a list of about a half dozen collectors and always set aside a few sets of tokens from each farmer to satisfy their needs. The tokens originally sold for 25 cents apiece, but later the price was raised to $1.00 for each token. The society now sells individual brass tokens for $4.00 each. The iron, aluminum, and fiber tokens sell for less as they are not used in making jewelry.

The Smiths made picker check jewelry for almost two decades (1967 - 1986), which probably consumed over 90% of all brass tokens donated to the society. Although they are not available to token collectors, they are worn with considerable pride by the relatives and descendants of the old farmers and many Anne Arundel residents with a great appreciation for their heritage.

Token Manufacturers

B. Hamilton Smith wrote in his diary on May 16, 1910, "Bought 1060 picker checks for $5.50." Farmers purchased their checks in the same part of the city where they obtained their pickers. Baltimore had several manufacturers of tokens from whom farmers could have purchased their checks. The most probable sources were Dorman's Stencil and Stamp Works, Francis X. Koehler, and A.K. Robins, as all three were known to have made tokens during the years picker checks were used.

Many Dorman tokens are readily identifiable by their distinctive die designs. The tokens of John Randolph Parker, "3rd Dist. Store," with the denominations surrounded by rays and the liberty head designs, are characteristic of the merchant tokens made by Dorman.

John F.W. Dorman is a familiar name to collectors of trade tokens. He started his business in 1866 and by 1879 he owned his own building located at 19 German Street. Dorman died in 1893, but his business continued to operate under his name.[2] The J.F.W. Dorman company made many tokens for merchants through the years. It is probable that many farmers bought their checks from Dorman as they continued in business well beyond the Anne Arundel truck farming era. An article in the ***Sunday Sun Magazine*** of December 5, 1954, discusses the checks of the J.F.W. Dorman Company:

> *Crop pickers' checks are still popular. These are marked with various quantity designations. As a field hand picks a quart, peck, bushel or some other quantity of fruits, berries or vegetables, he is given a check indicating it. At the end of the day, then, he exchanges his accumulated checks for cash. Pickers greatly prefer this system to other forms of tallying. They have been known to pick crops for a man who uses the check system, rather than pick at higher rates for a farmer who keeps tally with a pencil and pad.*

J.F.W. Dorman Company

By the 1930s most of the die-stamped tokens were used by the canneries on the Eastern Shore or other areas as the truck farms in Anne Arundel County were undergoing a rapid decline. Few survived much beyond World War II.

Another possible source of picker checks was Francis X. Koehler who operated a jewelry and die cutting business almost across the street from Dorman. F. X. Koehler was born and raised in Germany. He came to Baltimore in 1850 and started his business a year later. He struck tokens during the Civil War and continued in the token business until his retirement in 1885.[3]

One of the most probable sources of picker checks was A.K. Robins & Company located at 724-726 E. Pratt Street. They later relocated to 116 Market Place and, in the 1930s, were located in the block bounded by Lombard, Concord and West Falls Avenue. They also operated a foundry and machine shop in Cardiff, Maryland.

A.K. Robins & Co. was founded in 1855 and specialized in canning machinery and canner's supplies. Its 1909 catalog, listed a "full line of Brass, Fibre and Aluminum Checks for pickers, canners, pea-pickers, berry-pickers..." It listed five different sizes of round and octagonal checks for prices varying from 75 cents for 100 to $7.50 for 1,000. The checks came "stamped with initials and numbers." The company's 1920 catalog added square checks "With Sunk Initials and Numbers." Prices reflected the inflationary trends of World War I as they sold for $2.00 per 100 to $20.00 for 1,000. However, all prices were subject to discounts. In 1920, A.K. Robins & Co. added a complete line of brass and aluminum pay checks with raised letters and figures, which were offered in values of 1, 5, 10, 25, 50, and 100. The values were surrounded with rays and the costs were almost double the costs of the tokens with "sunk initials and numbers." Many canneries used the more expensive tokens, but the farmers, interested only in utility, preferred the sunk initials for obvious economic reasons.

It is interesting to note that the A.K. Robins tokens were furnished with the initials and values. From this we can assume that many farmers had their initials and values stamped by the manufacturer rather than stamp their own data with dies on the farm. Some farmers had their own dies and stamped their own initials on the checks as some checks are found counter-stamped with different initials and values.

Home-made tokens of Theodore H. Stinchcomb.

A few farmers made their own tokens. Theodore H. Stinchcomb on Mountain Road hand-cut his own tokens from copper, aluminum and bronze. His large, oblong tokens (41 x 23mm) are hand-stamped with initials and values. In addition, Mr. Stinchcomb punched dots, X's and V's on the reverse which corresponded with the values on the obverse, presumably to accommodate illiterate pickers.

Iron token of Plummer Arthur Clark.

The several known iron tokens also may have been handmade. They were cut by machine and stamped with initials and sometimes values. The iron tokens of Plummer Arthur Clark (PAC) exist in many different variations of values as Mr. Clark (or his children) seems to have gone wild stamping combinations on his tokens. It is difficult to tell what some of the values are.

Aluminum Tokens

In 1854 aluminum was worth $17 an ounce because aluminum was too expensive to separate from its natural ores. There was little incentive to devise new methods of extraction as there was very little market for aluminum. The only refinery in the world was in France, which had produced only about 60 pounds of the metal. The electrolytic process for refining aluminum, first introduced in the United States in 1888, drove the price down to several dollars a pound. By 1889 the French plant produced 50 pounds a day, but there was still no market for the metal. By 1910 aluminum was so cheap that new uses were being devised.[4] One of these was to make aluminum tokens which were now cheaper than other metals. Hence, we can conclude that if a picker check is made of aluminum, it can be dated no earlier than the first decade of the twentieth century. Several farmers used brass tokens initially, but converted to the cheaper aluminum tokens at a later date.[5]

Rarity

Every large collection inventoried revealed new types and varieties of picker checks which appear to be unique to the series. In some cases the number of tokens donated to the Historical Society is recorded under each farmer, but in most cases there are no records of the number of tokens used or the number that have survived. Farmers with large farms would purchase several thousand tokens at a time. Marvin R. Shipley remembered having about seventy-five pounds of his father's checks (L.R.S.), which could equate to well over 6,000 checks. Donations to the County Historical Society have included over 4,000 checks from a single farmer, while some donations consisted of only a few checks.

No attempt has been made to estimate the rarity of individual pickers' checks. In most cases rarities are unknown, but collectors must be aware that, even if several hundred tokens of a particular variety and value were donated to the Historical Society, most of them, as long as they are brass, are now found on bracelets or dangling from a lady's ears, leaving only a few available for collectors. A few tokens have surfaced outside the Historical Society and have escaped the captivity of jewelry, but these checks are not necessarily more plentiful for collectors as they were usually obtained from a farmer's relatives in small quantities.

Values

The value of a picker's check varies as to where it is purchased. The Ann Arrundell County Historical Society sells some tokens (along with the jewelry) in its Browse and Buy Shoppes. They now charge four dollars for each token as you pay extra for the historical value of the token. They also recognize the limited availability of the tokens and use the tokens as a means to raise funds to support the mission of the Society. From a coin dealer or an individual, pickers' checks can be purchased for as little as 25 cents. The tokens are often put in a dealer's junk box as most dealers know very little about the tokens. Rarity is usually unknown and has little bearing on the token's value.

Normally the tokens with higher denominations demand a greater price, and aluminum and pressed cardboard tokens can be purchased for less than the brass tokens.

Condition

Condition has little bearing on the value of a token as almost all tokens were used in the fields and passed through the hands of the farmers, rowbosses and pickers. Some larger value tokens saw little use and are found in a near uncirculated condition. Most brass tokens have been polished as was the case for all tokens used for jewelry. Some tokens were dug up or recovered from their graves in the farmer's old well. These are found in badly corroded condition and even attempts to polish them have proved futile. Most tokens from a particular farmer were placed into circulation at the same time and usually show the same degree of wear and do not exist in a range of conditions. If "nice" tokens are found, they may command a better price, but the difference is usually very slight. Basically condition is not a major factor in determining a token's worth. Regardless of the condition, the token still usually sells for "the going price."

Denominations

Thus far over 50 different denominations have been recorded on pickers' checks. With all the combinations of letters and values, there are over 130 known varieties and more will probably be discovered in the future. The following denominations have been recorded:

NV(No Value), ¼BP, ¼LP, ¼P, ¼BKT, ½, ½B, ½P, ½LP, ½BP, ½PK, ½BUS, ½SP, 1, 1QT, 1B, 1BP, 1LP, 1P, 1S, 1SP, 1¢, 1BKT, 1PT, 1R, 1PK, 1BUS, 1M, 1¼, 1½, 1½R, 2, 2¢, 2SP, 2BP, 2PT, 2R, 2LP, 2BUS, 2½, 2 ½B, 2½BUS, 2½P, 3, 3B, 4, 4¢, 4BKT, 5, 5¢, 5BP, 5LP, 5B, 5R, 5BUS, 5P, 5S, 5SP, 5½B, 6, 6B, 7, 7BP, 7½, 8, 8P, 8QTS, 9, 10, 10¢, 10R, 10B, 10BP, 10LP, 10P, 10S, 10SP, 12, 12½, 14BP, 15, 15S, 16, 20, 20¢, 20B, 20SP, 20P, 25, 25B, 25BP, 25R, 30, 30B, 30P, 37½, 40, 45, 50, 50B, 50BP, 50BUS, 50P, 55, 75, 90, 100, 100B, 100P, 100S, 100SP, 150, 200, 300, 400, 500, 600, 1000, 1200, 1500, 2000, 1.00, $1.00, 1.50, 3.00, ₴3.00, 5.00,"B", P, II, III, IIII,"1 BOX STRAWBERRIES," Star.

The letters are believed to stand for the following:

B - beans/bushel
BP - big peas/bushel peas
LP - large peas/peck
SP - small peas/peck
P - peas/peck
BKT - Basket
PK - peck
BUS - Bushel
QTS - Quarts
PT - pint
¢ - cent
S - strawberries(?)
R - raspberries(?)
M - melons(?)

Nomenclature

A system has been devised to standardize descriptions of the various types and varieties of picker checks. From this nomenclature individual tokens are identified by the farmers' initials, the size, the values, the material the token is made of, and design found on the border of the token.

The following abbreviations are used in the nomenclature:

Material: B - brass; A - aluminum; I - iron; Cu - copper; F - fiber (pressed cardboard).

Shapes:

Cl - Clover

D - Diamond

Ob - Oblong

Oc - Octagonal

R - Round

Sc - Scalloped

Sh - Shield

Sq - Square

Sw - Sawtoothed

Rims: The design found on the outer surface of the token.

br – beaded | fr - fiber | lr – line | pr - plain | rr - roped

Miscellaneous: lg - large | sm - small | NV - no value
r - red | b - black | v - varieties

Note: If two checks with the same farmer's initials and of the same value are made of the same material, are the same size and have the same rim design, but have different size numerals or different size letters, they are considered two different varieties, but are not listed as separate tokens.

Assumptions

Certain assumptions have been made to simplify the token's description. Variances from these assumptions will be noted in the token's description.

1. All tokens are incused (letters and values are hand stamped with steel dies by the farmer or manufacturer). If the token was manufactured with a die, it is noted in the description.
2. All tokens have **blank reverses** unless noted in the description.
3. All tokens are made of **brass** unless otherwise noted in the description.

All tokens are assumed to be **incused, brass with blank reverse** unless otherwise noted in the description. Using the abbreviations and the assumptions, a token would be described in this sequence:

Farmer - value - shape - size - border

U.W.B. 1 - R - 24mm - rr

Since it is not otherwise designated the token is brass, incused and has a blank reverse.
Another example:

H.O. - 5 - R - 24mm – pr - rF

Farmer - value - round - size - plain border - red fiber

In the catalog the farmer's initials are not used in the line description as each token is listed under the initials and the name of each farmer, if known. The value of the token would then be the first entry of each individual token listed under the general heading of the farmer's initials and identity, if known. The farmer's initials would only be used in the description if a token were listed separately.

Following is a catalog of all known Anne Arundel pickers' checks used on truck farms in Anne Arundel County, Maryland. Because of the difficulty in identifying similar checks from surrounding counties and the Eastern Shore, some of the unidentified checks may not be from Anne Arundel County. Please contact the author if you have further information on these checks, or if you have other checks not included in this inventory. Also, if you have conflicting information, please let the author know.

CHAPTER FIVE NOTES

1. Fletcher, S. W., *Strawberry Growing*, pp. 167-168.
2. Rulau, Russell, *United States Trade Tokens 1866-1889*, second edition (Iola, Wisconsin: Krause Publishing Company, 1998), p. 74.
3. Ibid, p. 76.
4. Ibid, p. 9.
5. Several families have reported that their father or grandfather had aluminum tokens made after using brass tokens for many years.

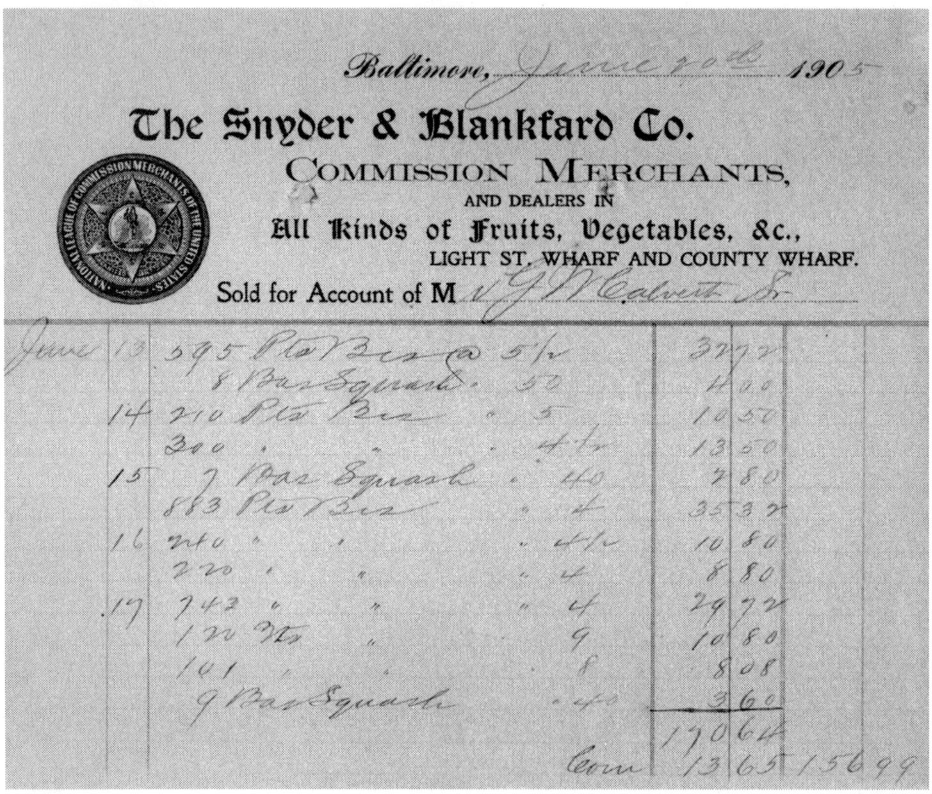

Baltimore, June 20th 1905

The Snyder & Blankfard Co.

COMMISSION MERCHANTS,

AND DEALERS IN

All Kinds of Fruits, Vegetables, &c.,

LIGHT ST. WHARF AND COUNTY WHARF.

Sold for Account of M J. W. Calvert Sr.

June	13	595 Pts Bes @ 5½	32 72	
		8 Bas Squash @ 50	4 00	
	14	210 Pts Bes @ 5	10 50	
		300 " " @ 4½	13 50	
	15	7 Bas Squash @ 40	2 80	
		883 Pts Bes @ 4	35 32	
	16	240 " " @ 4½	10 80	
		220 " " @ 4	8 80	
	17	743 " " @ 4	29 72	
		120 Qts " 9	10 80	
		101 " 8	8 08	
		9 Bas Squash @ 40	3 60	
			170 64	
		Com	13 65	156 99

Inventory

Inventory of
ANNE ARUNDEL COUNTY PICKER CHECKS

"A"

Albert — **James Albert Portaskewicz**
Millersville
(Changed name to James Albert)

C. H. A — **Carl H. Abend**
(Checks found in walls of an old home on Harness Creek Road.)
½Bus - R - br - Al
5Bus - R - br - Al

J. C. A. — **Josiah C. Armiger**, 1843 - 1928
High Point, Stoney Creek & Mt. Carmel, Bodkin Creek
(Lived at 900 William St., Baltimore)
1½ - R - 20mm - fr
1½ - Oc - 19mm - br
15 - Oc -
20 - Sc - 25mm - br
50 - R - 29mm - br (3 stars on obv.)

J. E. A — ?
NV - R

J. F. A. — **Joseph Frank Andrzejewski**, 1888 (Poland) - 1970
Dicus Mill Rd., Millersville
(Polish family started farming in 1919. Bought farm from H. C. Wigley, Sr. Used checks until 1956)
5p - Sc - 22mm - lr
20 - R - 21mm - pr
25p - Sc - 25mm - pr
500 - Oc - 23mm - pr

J. S. A. & Co. — **James S. Armiger & Co.**
Solley Road
(Checks were used by Clifford Bassford)
6 - Oc - br

J. W. A. ?
5 - Sc (found in a farm field)

R. H. A. **Richard Henry Arnold**
Mountain Road
2 - Oc - br

"B"

B ?
(Could be Bottomley checks as they were found with T.F.B. checks, but relatives do not recall "B" checks)
NV - R
NV - Sc
1½ - R - 21mm - fr
2 - Oc - 19mm - br
7½ - Cl - 28mm - pr
10 - R - 23mm - fr
15 - Sc - 26mm - br
20 - Oc - 27mm - br
25 - R - 23mm - fr
30 - Oc - 23mm - br
150 - R - 37mm - fr

BRUSH **Brush**
(Family recalls picker checks were used by their father and pickers recall being paid with pickers' checks)

A. L. B. **Arthur L. Boone**
Jacobsville
NV - R - 23mm - fr
NV - Sc - 26mm - br

B. B. B. **Basil B. Benson**
1 - R - Al
5 - R - Al
10 - R - Al
15 - R - Al
50 - R - Al
100 - R - Al

C. B. **Charles Boone**, 1858 - 1951
Jacobsville
2 - Oc - 22mm - br
3 - R - 19mm - pr
5 - R
15 - R
30 - Oc - 23mm - br

C. B. Jr. **Charles Boone, Jr.**

Jacobsville
1 - Oc (2 on rev.)
1½ - R
1½ - Oc - 19mm - br (2 or 10 on rev.)
5 - Sc
5 - Oc
10p – Oc – 22mm - br
10p - R - 20mm - fr
10 - Oc - 22mm – br (Some have "5" over the "0")
15 - Sc - 22mm - br (Some have "20" on rev.)
50p - Sc - 25mm - br (Some have "BJR" on rev.)
100p - Oc - 26mm - br
100 - R - 28mm - br

C & BRO. **Chairs & Brother** (Frank Chairs & Brother)
Marley P.O.
50 - Oc - 22mm - br ("JC" on rev.)
100 - Oc - 26mm - br ("JC" on rev.)

C. F. B. **Christian Frederick Ballman**, 1864 - 1949
Brooklyn
(Family recalls checks)

C. W. B. **Charles Wesley Brown**, 1850 - 1921
Mago Vista, Severna Park
(Used by the Schramm Brothers - rev. "S" or "S.B.")
1 - Sq - 23mm - pr(2var.)(rev. "S")
4 - R (In script)
5 - R - 24mm - fr
5 - Oc - 25mm - br (rev. "S.B.")
9 - Cl - 28mm - pr (rev. "S.B.")
10 - R - 25mm - fr
10 - Sc - 22mm - br (rev. "S")
20 - R - 24mm - fr (rev. "S")
45 - Sw - 24x28 - pr
90 - Sc - 25mm - br
100 - Oc - 26mm - br
500 - Sc - 26mm - br

G.B. **George Benning**
Galesville (Tomato Farm)
1 - R - 21mm - pr
5 - R - 21mm - pr
10 - R - 21mm - pr
25 - R - 21mm - pr
50 - R - 21mm - pr
100 - R - 21mm - pr
1000 - R - 21mm - pr

G. F. Beck 5 - R - 24mm - rr

G. H. B. **George H. Brice**
St. Margarets
5 - Oc

G. M. B. **G. Milton Benson**
1 - R - 34mm
5(rev) - Oc - 30mm
1 - Ob - 34x20mm

H. B. **Henry Brooks** (Black Farmer)
Furnace Branch Road, Cedar Hill (Father of Urias W. Brooks.)
1 - R - 21mm - br
2 - R - 21mm - br
3 - R - 21mm - br
4 - R - 21mm - br
5 - R - 21mm - br
6 - R - 21mm - br
7 - R - 21mm - br
7BP - Oc - 22mm - br
8 - R - 21mm - br
10 - R - 21mm - br
10SP - Sc - 22mm - br
14BP - Oc - 22mm - br
20SP - Sc - 22mm - br
100 - R - 21mm - br

H. C. B. **Henry Clay Bourke**, 1859 - 1945
Severna Park
NV - Sw - 24x28 - pr
NV - R
NV - Sc - 22mm - br
1 - R - tr
2 - R - 21mm - br
5 - R - 21mm - br

J. B. **Joseph Boyer** (?)
Severn
1 - R - 26mm
6 - R - 27mm

J. B. — **Johann (John) Blob**, 1889 - 1963
Dorsey Road (Brother of Max Blob)
1 - R - 22mm
50 - Oc - 23mm
100 - Oc - 23mm
500 - Oc - 23mm - br

J. B. — **John B.Blackowicz**, 1892 - 1960
Dicus Mill Road, near intersection Gambrills Road, Severn. (Nicknamed "Smitty", often called John Smith. Family remembers checks with initials, but no values.)

J. F. B. — **Johann Fredrich Bussey**, 1855 - 1924
Near Ice Rink opposite Benfield Road, Severn. (Fell off his wagon coming from Baltimore and died three weeks later from a severe concussion. His three sons also used his brass checks)

J. M. B. — **Joseph M. Brian**
Linthicum
5 - Oc

J. R. B. Jr. — **James Rufus Benson, Jr.,** 1893 - 1930
Mountain Road, Marley Creek
NV - Sc - 25mm - br
1s - R - 19mm - fr
5s - R - 21mm - br
10s - R - 24mm - fr
10p - Oc - 26mm - br
15s - R - 29mm - br

J. S. B. — **James S. Benson**
(Notice in ***Evening Capital*** - June 1910)

L. M. B. — **L. M. Boyer** (?)
Severn
50 - R

M. B. — **Mahlon A. Benson,** 1879 - 1951 (Also see M.A.B.)
Camp Meade Road, Hanover
10 - R - 24mm - tr

M. B. — **Maximilian Blob**, 1883 - 1969
8012 Max Blob Park Road, Jessup
(Now Blob's Park)
½B - R - 21mm - br
1 - R - 21mm - br
1B - R - 21mm br
2 - R - 21mm - br
2½B - R - 21mm - br
3 - R - 21mm - br
5 - Sc - 26mm - br
5B - Sc - 23mm
10 - Sc - 23mm
10B - Sc - 23mm
20 - Oc - 23mm - br
50 - Oc - 23mm - br
50B - Oc - 23mm

M. A. B. — **Mahlon A. Benson**
Camp Meade Road, Hanover
1 - R - 25mm ("1" on rev.)
5 - R - 25mm ("5" on rev.)
10 - R - 25mm ("10" on rev.)
10 - Sc - 23mm
15 - Sc - 23mm
100 - Oc - 25mm

O. & B. — ? (Could be a cannery token)
1000 – R – 32mm – fr

P. H. & B.
P. H. & Bro. — **Peter Hahn & Brother**, 1861 - 1941
Mountain Road, Pasadena
(Brother Edward Hahn, b. 1866)
NV - Sc - 28mm - br
1½ - Oc - 19mm - fr
10 - R - 20mm - fr
10 - R - 21mm - br
10 - D - 23mm - pr
15 - R - 21mm - br(2var)
15 - R - 23mm - fr
20 - Sc - 28mm - br
20 - Oc - 19mm
30 - Sc - 25mm
75 - R - 24mm - br
100 - Sc - 25mm - br
300 - Sc - 22mm - br
500 - R - 24mm - fr

R. T. B. — **Richard Tilghman Brice**
Broadneck
1 - Oc
3 - R
5 - R
10 - Sc
20 - Oc

Shriver Bothers - (see listing under "S")

T. B. **Thomas M. Beasley**
Severn
(Sold farm to J. C. Jacobs, Sr. Mentioned in newspaper article, June 1910. Family recalls tokens marked T. B.)

T. F. B. **Thomas F. Bottomley**
Ft. Smallwood Road, Magothy
(Came to the U.S. from England as an infant. Son-in-law of R. W. Chard. Checks were also used by his son, Robert W.Bottomley, the originator of the Bottomley cantaloupe. Photos from Lewis Hines were taken on the Bottomley farm in 1909. Some of the 7½ checks have "10" stamped on the reverse.)
NV - R (rev. "RWC 1½")
½ - Sc
1½ - Sc - 22mm - br(2var)
1½ - R
2 - R - 20mm - fr
7½ - R - 23mm - fr(2var)
7½ - R - 20mm - fr
7½ - Oc - 22mm - br(2var)
10 - R - 23mm - fr
15 - R - 20mm - fr
15 - R - 23mm - fr
20 - Oc - 22mm - br
30 - Sc - 25mm - br(2var)
75 - R - 29mm - br(2var)

U. W. B. **Ulysses W. Brooks**
Furnace Branch Road, Glen Burnie
(Black Farmer, son of Henry Brooks. Name could be Urias, 1877 - 1964. Farmed 50 acres on Furnace Branch Road and 10 acres on Ritchie Highway. He was a pioneer settler of this community.)
1P - R - 23mm - fr
1 - R - 24mm - rr
2 - R - 24mm - rr
5P - R - 23mm - fr
5 - R - 24mm - rr
10 - R - 24mm - rr
15 - R - 24mm - rr(2var)
500 - R - 24mm - rr

W. B. **William Brossure**
Crownsville
5 - R - 21mm - fr

W. L. B. **Will L. Bussey**
Severn (Family recalls checks)

W. O. B. **Rev. William O. Bussey**
Severn (Family recalls checks)

W. P. B. **William P. Baldwin**
St. Margarets
1 - R - 24mm

"C"

C & Bro **Chair & Brothers** (Listed under B's)

B. F. C. **Benjamin Franklin Chaney, Sr.**, 1880 - 1949
St. Stephens Church Road, Gambrills
(Son remembers round, brass tokens of the values: 5, 10, 20, 25. All had holes in them as they were kept on a wire ring)

B. F. C. **Benjamin F. Clark**
Reece Road, Severn
(Family recalls tokens)

B. S. C. **Basil Smith Cromwell**, 1856 - 1926
Welham Station, Ferndale
(Later purchased John A. Wagner's farm on Old Telegraph Road. All checks are stamped on the reverse "E.L.C." for Emory Luther Cromwell, son of Basil Cromwell.)
1 - R - 20mm - fr
5 - R - 20mm - fr
10 - R - 23mm - fr
10P - R - 23mm - fr

C. C. **Charles Clark**, 1858 - 1935
Severn
(Original farm on Reece Road in two locations. Checks were used for strawberries only.)
1 - R - 20mm
5 - R - 20mm
10 - R - 21mm
20 - R - 21mm
50 - Sc - 22mm
100 - Sc - 25mm
500 - Oc - 23mm

C. H. C. **Charles Hancock Calvert, Sr.**, 1881 - 1972
On Deep Creek, Bodkin, between Popular Ridge Rd. and Bayside Beach Road.
(Son of George W. Calvert and father of James B. Calvert)
Star - R - 26mm
Star - R - 29mm(holed)

C. W. C. **C. W. Cromwell**
Cock's Creek, Ft. Smallwood Rd.
(Family recalls tokens)

E. C. **Eugene Clark**
Gambrills, now Crofton
(Farm called "Linthicum Walks," now the home of the Anne Arundel Heritage, Inc., listed on the National Register in 1984. Family recalls checks.)

E. D. C. **(Cromwell) ?**
(Token found in a well with BSC tokens)
8 Qts - R - 21mm – pr - rF

E. L. C. **Emory L. Cromwell**
Ferndale
(Son of Basil Cromwell. "ELC" stamped on reverse of BSC checks)
NV - R

F. C. **Franklin Chairs**
near Ft. Smallwood Rd, Pasadena
(Most checks have a star on the reverse)
6 - Sc - 22mm - br(no stars)
6 - R - 20mm - br
10 - Sc - 22mm - br(2var)
20 - R - 23mm - br
50 - Oc - 22mm - br
100 - Sc - 22mm

F. P. C. **Frank P. Chairs**
Pasadena
6 - R - 20mm - br

F. W. C. **Frederick W. Cook**
NV - Sc - 25mm - br
NV - R - 21mm - br

G. T. C. **George T. Cromwell**
(Another son of Basil Cromwell.)
10 - R
20sp - Oc
50sp - Sc - 25mm - br (G.T.C. on rev. found in a field)

G. W. C. **George W. Calvert**
(Pickercheck box with token values written on each compartment, in possession of author. Values of 1½, 2, 10, 15, 20)

H. C. **Harry A. Cook**, 1858 - 1934
Bayside Beach Road, Pasadena
1 - R - fe
5 - R - 24mm - fr -

H. A. C. **Henry Alfred Cook**, 1858 - 1934
1817 Bayside Beach Road, Pasadena
(Used tin tokens which were sometimes redeemed at Captain Henry Wickhart's Store on Popular Ridge.)

H. D. C. **Henry Deville Cook**, 1865 - 1925
Paradise Beach Ave., Pasadena
1½ - R - 21mm
4 - Oc - 22mm
5 - Sc - 25mm
10 - R - 19mm - fr
20 - Oc - 22mm

H. M. C. ?
3 - R - rF
4 - Oc

J. C. **J. Chairs**
Mountain Road, Pasadena
(Son of Frank Chairs. "MC"and"C & Bro" tokens have "JC"on the reverse)

NV - R - 20mm - fr(rev"12½")
NV - R - 24mm - rr
6 - Sc - 22mm - br
10 - R - 21mm - br
12 - R

J. H. C. **James Henry Crouse**
Gambrills Road, Severn, next to Wilson Disney.
(Brother of William T. Crouse. Family recals tokens, 1, 5, 15 and others, used only for strawberries.)

J. H. C. **John Henry Carroll**, 1839 - 1907
New Cut Road
(Black Farmer born a free man. Mother was a slave and his wife was a slave.)
1 - R - 20mm - fr
1 - R - 21mm - br
3 - R - 23mm - fr
3 - R - 24mm - rr
5 - Sc - 22mm - br(2var)
15 - Sc - 26mm - br
50 - Oc - 22mm - br
50 - Oc - 26mm - br
100 - Sc - 28mm - br
150 - R - 29mm - br

J. M. C. **Jefferson Monroe Cook**, 1841 - 1925
Mountain Road near Pinehurst.
(Sent provisions by boat to the Confederacy.)
5 - Oc - br

J. S. C. **John Sewell Clark, Jr.**, 1878 - 1949
Clark Road, Severn
(Family recalls tokens)

J. T. C. **J. T. Clark**
Jessup
5 - R

J. W. C. **John Wesley Clark**, 1832 - 1908
Reece Road, Severn, (now 1020 Reece Road)
(Father of Asbury W. Clark. Farmed on "Trusty Friend" purchased in 1858. Some tokens have "WS" counterstamped for William Smith who purchased the tokens at auction)
NV - R - 23mm - fr(rev "no 216", holed)
½p - R - 20mm - fr
1 - R - 20mm - fr(2var)
1p - R - 20mm - fr
5 - R - 20mm - fr(2var)
5p - R - 20mm - fr
10 - R - 20mm - fr
10p - R - 20mm - fr
20 - Sc - 22mm - br
20p - R - 20mm - fr
25p - R - 21mm - fr
50 - Sc - 21mm
50p - R - 20mm - fr
100 - Sc - 22mm - br
100 - Sc - 25mm

M. C. **Monroe Cook**
Pasadena
10 - Sc - 25mm - br(rev "J.C.")
20 - R - 23mm - fr (rev "J.C.")
50 - R - (M.C. c/s over "J.C.")
50 - Sc - (M.C. c/s over "J.C.")
100 - Sc - (M.C. c/s over "J.C.")

P. C. **Plummer Clark**
Rt. 175 & Rockenbach Road, Odenton
(Plummer Clark used both P.C. and P.A.C. checks.)
150 - R - 51mm – pr - I

P. A. C. **Plummer Arthur Clark**, 1858 - 1944
Rockenbach Road & Md. Route 175
(Iron checks found with many different varieties, punched "PC" and "P.A.C." or both. Son used these checks until June 1917.)
I - R - 35mm – pr-I(2var)
I - R - 51mm – pr-I
II - R - 51mm – pr-I
III - R - 51mm – pr-I
IIII - R - 51mm - pr-I
5 - R - 35mm – pr-I
10 - R - 35mm – pr-I(2var)
15 - R - 35mm – pr-I(2var)
50 - R - 51mm – pr-I(4var)
150 - R - 51mm – pr-I(4var)
150 - 50 - R - 51mm – pr-I

R. N. C. **Richard N. Cromwell**
½p
100

R. S. C. **Richard Stanley Cole**
Friendship Airport area
(Checks were sold for scrap metal.)
2 - Oc - 22mm - br

R. W. C. **Robert Wesley Chard** 1829 - 1893
Ft. Smallwood Road, Magothy
(Father-in-law of Thomas Bottomley)
NV - R - 29mm - fr
1½ - R - 23mm(some have T.F.B. on rev.)
1½ - R - 21mm
7½ - Sq - 23x26mm - pr(corners hand cliped)
15 - R - 20mm - fr
30 - R - 23mm - fr
75 - R - 29mm - fr(2var)
1.50 - R - 35mm - pr

S. A. C. ?
NV - R
5 (on rev.) - R - 24mm

S. S. C. ?
NV - R - 20mm - br ("5" in wreath)

S. T. C. Jr. **Sedwick Thomas Cromwell Jr.**
Wellhams
(Father STC (1800-1843) was an early truck farmer.)
1 - Sc - 25mm - br

S. T. C. of J. — **Sedwick Thomas Cromwell of John**, 1862 - 1950
Crain Highway, Glen Burnie
(The "p" on the ½, 5, and 10 checks is above the value.)

½p - R - 20mm - fr
½BUS - Sc - 28mm - pr
1 - R - 20mm - fr
1BUS - D - 23mm - pr
2½ BUS - Sw - 24x28 - pr
5 - Sc - 22mm - br
5p - Sc - 22mm - br
5BUS - Cl - 28mm - pr
10 - Oc - 22mm - br
10p - Oc - 22mm - br
100 - Sc - 25mm - br

T. A. C. — ?
½P - R - 23mm - fr (found in a field
2½P - R
3 - R
100 - Sc

T. M. C. — **Thomas M. Cole**
Harmans
(Checks first used by TMC from 1880 to 1908, sold to Samuel Vinton Clark, 1853-1922, who used them from 1908 til 1918 when they were used by his son, Gilbert H. Clark, 1892-1974, & his brothers until 1945. Mr. Samuel V. Clark was relocated when Ft. Meade was established.)

½p - R - 20mm - br
½sp - R - 24mm - fr
1p - R - 20mm - fr
1sp- R - 24mm - fr
1 - R - 23mm - fr
1 - R - 20mm - fr(2var)
2 - R - 23mm - fr
2 - R - 20mm - fr
2sp - R - 24mm - fr
5 - R - 20mm - fr
5p - R - 20mm - fr
5sp -R - 24mm - fr
10 - R - 20mm - fr
10p - R - 20mm - fr
10sp - R - 24mm - fr
50 - R - 23mm - fr
50p - R - 23mm - fr
100 - R - 23mm - fr (some C/S "100" on rev.)
100p - R - 23mm - fr

T. W. C. — **Thomas W. Cole**
Harmans
1 - R - 23mm - fr
5 - R - 23mm - fr

V. B. C. — **Van Buren Carter** (?)

W.C. — **William Cromwell**
(Reported to have used checks)

W. H. C. **William Hinkle Chairs**, 1868 - 1942
Lake Shore, Pasadena
(Family recalls checks)

W. H. C. **William Handy Clark**, 1865 - 1929
Severn
(Son of Benjamin F. Clark, family recalls tokens)

W. H. C. **William Henry Clark**, 1841 - 1904
Severn
(Used Joshua Owens Picker checks.)
10 - R

W. L. C. **William Landy Cook**, 1886 - 1968
Mountain Road, (near Gibson Island)
(Started farming in 1908. 1,459 checks donated to the AACHS)
1½ - R - 22mm - br
½ - R - 21mm - pr
2 - R -21mm - pr
7½ - Sc - 19mm - pr
10 - R - 24mm - fr

15 - Oc - 23mm - pr
15 - Oc - 25mm - pr
20 - R - 24mm - rr
50 - R - 29mm - br
100 - Cl - 28mm - pr
1.00 - R - 28mm - pr

W. S. C. **William Shipley Crisp**, 1856 - 1917
Brooklyn
(Believed to be the grandson of the early strawberry farmer William Crisp.)
1 - Cl
1 - R - 20mm - fr
5 - Cl
5 - Oc - 19mm - br
10 - Sc - 22mm - br
1.00 - Cl
100 - Sc - 25mm - br

W. T. C. **William Thomas Crouse**, 1877 - 1961
Gambrills Road, Severn
(Brother of James Henry Crouse, Sr. Checks are die struck with rays and values on the reverse)
1 - R - 20mm - lr
10 - R - 20mm - lr
20 - R - 20mm - lr
50 - R - 20mm - lr

"D"

D — ?
NV - Oc
2 - Oc - 25mm - br

DICKLER — **Dickler**
(One check found in a farm field)
5 - R - 22mm

D. Bros. — **John and Jim Disney**
Fort Meade, MD
(Reported to have used checks)

A. J. D. — **Andrew J. Disney**
Stony Run Road & Elkridge Landing Road, Hanover
(280 checks donated to AAC Hist.Soc.)

1 - R - 20mm - fr	10 - Oc - 22mm - br
2 - Sc - 22mm - br	10 - Sc - 25mm - br
4 - R - 23mm - fr	20 - R - 23mm - fr
5 - R - 23mm - fr	20 - Oc - 22mm - br
	50 - Sc - 25mm - br

C. M. D. — **C. M. Duvall**, 1857 - 1924
Chesterfield
(Reported to have used checks)

D. F. D. — **David F. Donaldson, Sr.**, Died 1937
Donaldson Ave., Severn
(Reported to have used checks in the 1890s. Checks were later used by his sons David, Jr. and Paul.)

G. D. — **George Downs**
Now within the BWI area
(Relatives recall tokens and report that Mr. Downs buried about 50 lbs. of checks on his old farm.)

J. D. — **John Downs**
(mentioned in a June 1910 article)
3 - R - 24mm (hole in center)
5 - Ob - 33x19mm (hole in center)

J. G. D.	**J. G. Downs** Ferndale NV - Oc - 22mm 19 - R - 22mm
J. T. D.	**John Theodore Downs** Camp Meade Road, Ferndale NV - Oc - 22mm 10 - R - 22mm 20 - Sc - 22mm
M. D.	**Mahlon Disney**, 1871 - 1954 Mountain Road (Original farm was in Severn. Bought home and farm from August Carl Schmidt on Mt. Road in 1902. Son, Wilson Disney, farmed next to the Crouses.) 1 - Sc - 22mm 1L - R - 24mm 10B - R - 24mm 20 - R - 22mm 20 - Sc - 22mm 20 - R - 24mm 50 - Sc - 22mm 100 - Sc - 22mm
R. D.	**Rufus Day** Farmed in 8100 block of New Cut Rd., Severn. (Family recalls tokens)
S. E. D.	? (Tokens found on a farm on Severn Road) s10 - R - 22mm 20 - Sc - 28mm
T. D.	? 10 - R 100 - R
U. G. D.	? 5 - Oc - 25mm - br (Found in field by metal detector) 10 - Oc - 25mm - br
W. D.	**Wilson Disney**, 1827 - 1889 Gambrills Road, Severn (Also farmed on Severn Road, south of Ft. Meade. Believed to have used picker checks.)

W. D. **William Clarence Dotson**, 1841 - 1927
Furnace Branch Road, Glen Burnie
(Black farmer born a slave. Civil War veteran. After gaining his freedom, he changed his name to Dotson from Dodson to rid his image of slavery. Tokens also used by sons Daniel C. Dotson and Richard James Dotson. Sons stamped an "RR" on the reverse for the tokens they used.)

NV - Sc - 22mm
NV - Oc - 22mm
1B.P. - Sw - 24x28mm - pr
1 - R - 20mm - fr(2var)
2 - R - 20mm - fr(2var)
3 - R - 20mm - fr(2var)
5 - Sc - 22mm - br
10 - Oc - 22mm - br
20 - R - 23mm - fr
50 - Oc - 19mm - pr
100 - Sc - 25mm - br

W. P. D. **William Paul Disney**
Severn

1 - Oc - 19mm - fr
5 - R - 20mm - fr
10 - R - 21mm - br
20 - Sc - 22mm - br
30 - Oc - 23mm - br
40 - R - 24mm - rr
50 - Cl - 28mm - pr
100 - Oc - 25mm - br

W. T. D. **William Theodore Downs**
Hollins Ferry Road, Ferndale
(Relatives recall tokens)

"E"

G. W. E. ?
5 - R - 25mm

J. F. E. **John F. Ellison**, 1840 - 1908
Gibson Island
(Nephew, John A. Ellison, 1891 - 1981, farmed on Mountain Road near Mt. Carmel, used his uncle's tokens. Farm was still in operation in 1999, operated by granddaughter and husband, Cindy & Rick Holt. They also operate a roadside stand.)

1½ - R
3 - Sc
7½- R - 20mm
7½ - R - 23mm
10 - R (2 var.)
15 - R
20 - Oc
30 - R
45 - R - 23mm
100 - R
100 - Sc

O. E. **O. Elzey**
Race Road, Dorsey
(Family recalls tokens)

W. E. ?
Old Oak Road,(Knopp Farm)
10 - R (found in a farm field)
25 - Oc - 23mm - br
50 - Oc - 23mm - br

"F"

D. F. ?
5 - R

F. F. **Frank Fields**, 1871 – 1959
Farmed near Piney Run in Fourth District
(Frank Fields was a lawyer for the General Accounting Office in Washington. He was involved in the Panama Canal Project. He was a part-time farmer and commuted to work on the B & O Railroad. Other tokens were used.)
3 – R – 21mm – br
20 – R – 24mm - lr

F. S. F. **Frank S. Fisher**
Piney Grove area, Elvaton
1 - R - 21mm - pr
2 - Oc - 19mm - br
3 - Sc - 22mm - br
5 - Sc - 25mm - br
25 - Oc - 23mm - br
50 - Sc - 29mm - pr
100 - Sc - 29mm - br

G. F. F. **George F. Farring**
Brooklyn
(Family recalls tokens)

H. F. ?
(One token found in a field)
NV - R - 25mm (H.F. on rev., F. & C. with H. above & below on obverse.)

J. F. ?
5 - R
10w - Sc

J. G. F. — **John G. Friedhofer**, 1871 - 1931
Ft. Meade
1½ - R - 21mm - br   20 - Cl - 28mm - pr
20 - R - 21mm - br   500 - Oc - 23mm - br

J. H. F. Jr. — **Joseph Henry Frantum, Jr.**
NV - Sc - 22mm

H. C. F. — **H. C. Foster**
Mountain Road
2 - R

R. T. F. — **Ricard Thomas Ford**
Shipley Station, Camp Meade Road, Linthicum
(1,681 checks donated to AAC Hist. Soc.)
1 - R - 21mm - br   10 - Sw - 24x28mm - pr
2 - Sc - 22mm - br   15 - Cl - 28mm - pr
4 - Oc - 22mm - br   20 - Sc - 28mm - pr
8 - Oc - 22mm - br

W. F. — **William Fairbanks**
Farmed at Elkridge Landing Rd. & Balt.-Wash. Expressway, Linthicum.
(Family recalls tokens)

"G"

B. G. — **Benjamin Garner**
(Ad in **Evening Capital**, May 26, 1894: "I HEREBY FOREWARN all person from receiving my Checks." No checks known to exist)

B. G. — **Boone Griffith**
Jacobsville
(Family recalls tokens)

C. G. — **Columbus Griffith**
1.00
5.00

C.B.G. — **Columbus Boone Griffith**
7700 block of WB&A Road, Severn
(Family recalls tokens)

C. M. G. — **Charles Milton Green**, 1869 - 1937
Benfield Road, Millersville
(Also used paper chits signed by C.M.Green, made out to "Ben, Annie, Frank, Katie, and Rosie," dated 1918-1920)
1 - R - 20mm - fr
3 - Sc - 22mm - br
5 - Oc - 19mm - fr
10 - Cl - 28mm - pr
15 - R - 23mm
50 - Oc - 22mm - br
100 - Sc - 25mm - br

F. A. G. — **Francis Asbury Griffith** (?)
Farmed at 5th & Delmar Ave., Glen Burnie.
(He was called "Asbury")
500 - R - 23mm - fr

G. A. G. — ?
7½ - Sq - 25mm - pr(some handcut into Oc)
15 - R
30 - R

H. G. — **Herman Gunther** 1877 - 1951
Farmed on North Shore Road, Pasadena.
(Family recalls checks.)

H. E. G. — **Henry Emil Gertz** 1867 - 1946
St. Stephens Church Road, Chesterfield
(Born in Germany. Wife threw out most metal checks. HEG also used cardboard chits, "Peas" - Sq - 27mm with "Gertz" on the reverse; used for one bushel of peas.)
1 - Oc - 23mm – pr - rF
5 - Sc - 22mm – pr - rF
10 - Sc - 26mm – pr - rF
20 - Cl - 29mm – pr - rF
100 - R – pr - rF
500 - Oc - 29mm – pr - rF

H. E. G. — **H. E. Green**(?)
Dicus Mill Road
1 Oc - 23mm - br
5 - Oc - 23mm - br
20 - Oc - 23mm - br

J. H. G. **John Henry Gischell**, 1867 - 1953
Andover
(Checks were used by the Wengerts of Glen Burnie.1,569 checks were donated to the AACHS with metal box. Gischell was a part-time farmer who farmed in 1907, where the old Andover High School was located.)

½ - R - 20mm - fr
½ - R - 35mm - pr - I
1 - R - 20mm - fr
2 - R - 20mm - fr
4 - R - 20mm - fr
8 - R - 20mm - fr
10 - R - 20mm - fr
50 - R - 20mm - fr
100 - R - 20mm - fr

K. G. **K. Grienison** (?)
(One check found on the grounds of the Benson-Hammond House by John Stoll.)
5 - R - 23mm - fr

M. G. **Minnie Gruhn**
Truck House Road
(Family recalls tokens)

P. X. G. ?
10 - R - 25mm

R. W. G. **Robert Wesley Griffith**, 1849 - 1925
4th Ave., Glen Burnie
(Tokens originally had initials on the obverse and the values on the reverse. Many variations exist in combinations of values counterstamped on the checks. Many have been c/s to change the original value. Most Sw tokens have been changed from "15" to "1" by stamping over the "5". Robert Elmer Griffith, Old Stage Rd., may have also used these checks.)
NV - Oc - 23mm - br(rev "100")
NV - R - 23mm - fr(rev "001" or "10")

½B.P. - Sc - 22mm - br
½L.P. - Sc - 22mm - br
1B.P. - Sc - 22mm - br
1L.P. - Sc - 22mm - br
1 - Sc - 22mm - br
1 - R - 23mm - fr
1 - Oc - 22mm - br
1 - Sw - 24x28mm - pr
5B.P, - Sc - 22mm - br
5L.P. - Sc - 22mm - br
5 - R - 23mm - fr
5 - Oc - 22mm - br
5 - Sw - 24x28mm - pr
10B.P. - Sc - 22mm - br
10L.P. - Sc - 22mm - br
10 - Oc - 22mm - br
10 - Sw - 24x28mm - pr
10 - R - 23mm - fr
15 - Sw - 24x28mm - pr
100 - R - 23mm - fr
100 - Oc - 22mm - br

S. L. G. **Samuel L. Gaylor**
Ridge Road
1 - R - 21mm - fr
8p - Oc

T. J. G. **Thomas Jefferson Gardner**, 1863 - 1955
Old Oak Road, Severn. Later sold farm to Knopp family.
(Family recalls tokens)

W. G. **William Gunther**, 1911 - 1991
Old Annapolis Road, Marley

1 - R - 26mm - pr - Al	10 - R - 22mm - pr - Al
4 - Sc - 25mm - pr - Al	1.00 - Oc - 26mm - pr - Al
5 - Sq - 24mm - pr - Al	

W. M. G. **William M. Griffith**
Thought to have farmed in Severn.

4 - Oc - 23mm - br	15 - Oc - 23mm - br
5 - Oc - 23mm - br	20 - Oc - 23mm - br
8 - Oc - 23mm - br	
10 - Oc - 23mm - br	

W. T. G. **W. T. Gaylor** (?)
Odenton
(Checks used with H.D.C.)

½B - R - fr(2var)	½p - Oc - 23mm
1s - R - 21mm - br	2s - R - 21mm
5B - Sq - 21mm - pr	3s - R - 21mm
100S - R - 24mm - rr	10s - R - 21mm

"H"

A. H. **Arthur Helmer** ?
Farmed on Route 2 near Horn & Horn restaurant.
(Token found in field by metal detector.)
1½ - R - 18mm

A. C. Houseman A. C. Houseman
10 – R – 25mm – pr - I

A. S. H. **Albert Sydney Hammond**, 1863 - 1932
Marley Creek
(Named after Albert Sydney Johnston, CSA General.
Brother of Rezin Howard Hammond)
½L.P. - Sw - 24x28 - pr
1B - Sc - 25mm - br
50 - R - 24mm - br

B. H. **Bernard Helmer**
Glen Burnie
(Father of Arthur Helmer, born in Germany. Farmed in the
1600 block of Crain Highway. Bought farm around 1890.)
7½ - R - 21mm
50 - Oc - 23mm
100 - Oc - 22mm

C. A. H. ? (Not confirmed Anne Arundel County tokens)
½ - R - 20mm
1 - Oc - 23mm
2 - R - 23mm
5 - Oc - 23mm
7 - Oc - 23mm
8 - Oc - 23mm
100 -

C & R. T. H. **Clarence & Richardson Thomas Hammond**
1882-1928, 1883-1952
Old Ordnance Road
(Sons of John Thomas Hammond and brothers of Rezin Howard
Hammond. Some checks have C & RTH on both sides)
1 - Sc - 25mm - br
5 - Sc - 25mm - br
10 - Sc - 25mm - br
15 - Sc - 25mm - br
20 - Sc - 25mm - br
25 - Sc - 25mm - br

C. H. H. **Charles Henry Hines**, 1864 - 1947
Belle Grove Road
1BKT - R

C. H. H. **Charles Hawkins Hodges**
(Reported to have used checks.)

D. H. ?
10 - Sc - 22mm - br

D. L. H. ?
1/4BKT - R

E. H. H. **Egbert Holmes Hawkins**, 1893 - 1981
Harmans
(Son of John Sterling Hawkins,Sr.)
½ - R - 21mm - pr
1 - R - 21mm - pr
2 - R - 21mm - pr
5 - R - 21mm - pr
10 - R - 21mm - pr

E. J. H. **Edward James Hines, III**, 1822 - 1926
Bodkin Creek
(oddities exist with various stampings)
NV - Oc - 19mm - fr
NV - R - 23mm - fr
5 - R - 20mm - fr
5 - R - 23mm - fr
10 - Sc - 25mm - br

E. J. H. Jr. **Edward James Hines, Jr.**
Bodkin Creek
(son of E.J.H.)
10 - R - 23mm - fr
10 - Sc - 25mm - br

F. L. H. **Frank L. Hancock**
Farmed "Patapsco Plains" near Solly P.O.
(Frank farmed with his two brothers, Lemuel B. and Harvey L. Hancock. Their 325 acre farm is described in June 14, 1910 issue of ***The Sun*** newspaper. They made $2,000 on strawberries in one day in 1902. The article mentions that the pickers are paid in metal checks which are redeemable only on one specified date at the end of the picking season. None of their checks have been reported.)

H. H. **Herbert Hammond**, 1875 - 1953
Marley Creek
(Brother of Howard Hammond, father of Anna Hammond Coe. Son of J.T.H. Moved to Riesterstown in 1918.)
1 - R - 20mm - fr

H. R. H. **Rezin Howard Hammond**, 1864 - 1928
Hammonds Ferry Road, Linthicum
(Farmed at the Benson-Hammond House, "Cedar Farm")

1pt. - R - 20mm - fr
1B.P. - R - 20mm - fr
2S.P. - R - 24mm - fr
2B.P. - Sc - 22mm - br
2pt. - Sc - 22mm - br
25B.P. - Sq - 23mm - pr
50 - Oc - 22mm - br
100 - R - - 23mm - fr

J. H. **James Hopkins**
(Family recalls tokens)

J. H. **James Hawkins**
(Brother of Richard Walter Hawkins)

J. H. **John Hawkins**
2 - R
5 - R - 21mm - br

J. A. H. **James Asbury Hancock**, 1837 - 1903
Hancock's Resolution, Bayside Beach
(Iron tokens believed to be the earliest tokens used.)
NV - R - 35mm - pr - I

J. B. H. ?
5 - Sc - 23mm - br
10 - R - 21mm - br
20 - Oc - 23mm - br

J. G. H. **J. G. Hollins**
Annapolis
(Reported to have used checks.)

J. H. H. **John Henry Hancock**, 1872 - 1962
Maynards
(Son of Henry Alfred Hancock, half brother of James A. Hancock.)
NV - R

J. L. H. **Joshua Lackland Higgins**, 1839 - 1904
Gambrills - USNA Dairy Farm
(Farm, known as "Fairview," was sold to the U. S. Naval Academy in 1913. Operated until 1998 as Anne Arundel County's last dairy farm.)

1 - Oc - 19mm - pr
3 - R - 21mm - br
5 - Sc - 23mm - br
10 - Oc - 23mm - br
20 - R - 24mm - br
50 - Cl - 28mm - br
100 - Sc - 26mm - br

J. S. H. **John Sterling Hawkins, Sr.**, 1865 - 1894
Welhams
(Son-in-law of Hezron N. Kelly)

1 - R - 20mm - fr
2 - Sc - 22mm - br
4 - R - 22mm - fr
10 - Sc - 25mm - br
15 - Cl - 28mm - pr
20 - Oc - 22mm - br
100 - R - 28mm - br
500 - Oc - 26mm - br

J. T. H. **John Thomas Hammond**, 1839 - 1909
"Old" Ordnance Road
(Son, also John Thomas Hammond, 1872 - 1940, used the same tokens.)

M. H. **Marmaduke Hamilton**, 1858 - ?

1 - R - 19mm - pr - rF
½ - R - 21mm

O. H. ?
(initials O.H. are found on the reverse of A.C.W. checks)

P. H. ?
50 - R - 23mm

P. H. & B. **Peter Hahn & Brothers**
(Listed under B's)

R. W. H. **Richard Walter Hawkins**
(Brother of James Hawkins. Used checks)

R. W. H. **Rezin W. Hammond**
5th District
(Reported to have used checks)

S. P. H. **S. P. Hamilton**
(Reported to have used checks)

W. S. H. **William Shipley Hammond**, 1866 - 1924
Ordnance
(1,246 checks donated to the AACHS by DorothyHammond Smith in 1987. Number donated in parenthesis)

1 - Sc - 25mm - br(224)
2 - Sc - 25mm - br(259)
3 - Sc 25mm - br(161)
5 - Sc - 25mm - br(62)
10 - Sc - 25mm - br(40)
§10§ - Sc - 25mm - br(44)
15 - Sc - 25mm - br(40)
§15§ - Sc 25mm - br(98)
25 - Sc - 25mm - br(38)
50 - Sc - 25mm - br(41)
100 - Sc - 25mm - br(100)
500 - Sc - 25mm - br(139)

W. T. H. ?
1½ - R - 21mm – pr - rF
7½ - R - 21mm – pr - rF

W. W. H. ?
3B - R - 23mm - fr
6B - R - 23mm - fr
50B - R - 23mm - fr

"J"

C. N. J. **Cyrus Newton Joyce**
Joyce Lane, Arnold
(Farmed from 1870 to 1918. Shipped produce to Baltimore by steamboat from wharf at Joyce Lane. Reported to have used checks)

E. J. **Enos George Jeffrey**, 1866 - 1926
Dorsey Road, Hanover
(Reported to have used checks)

E. G. J. **Elmer G. Jubb**
Near Long Point on the Magothy, Pasadena
(Reported to have used checks)

F. J. **Fletcher Joyce**, 1915 - 1940
Severn Crossroads, Indian Landing Road, Millersville
(All tokens are red Fiber)
1 - R - 24mm – fr - rF
1½ - R - 24mm - fr - rF(rays)(1½ on rev)
5 - R - 24mm - fr - rF

F. J. — **Frank Jeffrey**
BWI area

1 - R
15 - Sc
20 - Sc
50 - Oc
100 - Oc
500 - Oc

G. A. J. — **George Albert Jenkins**, 1892 - 1960
Mt. Carmel on-the-Bodkin, Pasadena
(Farmed "Bob's Hill" on Forest Glen Road and managed two other farms, Piggery and Heath farms in the same area. Checks are corroded as they were recovered from a well. Son of Gustavus G. Jenkins)

1BKT - R - 21mm - pr
4BKT - R - 21mm - pr
(1½ and 3 were also used.)

G. C. J. — **G. C. Jenkins**
Tick Neck Road, Pasadena

6 - R - 24mm - fr
10 - Sc - 22mm - br

G. G. J. — **Gustavius G. Jenkins**, 1847 - 1915
Stoney Creek, Jacobsville

5 - R - 23mm - fr
5 - R - 24mm - pr(rev "not transferable")
6 - R - 24mm - fr
10 - Sc - 22mm - br(2 var)
20 - Oc - 22mm - br(2 var)

J. A. J. — **James Albert Jenkins**, 1863 - 1933
Jenkins Road, Rivera Beach
(Bought the old W. J. Lockerman farm on 8500 block of Jenkins Road. Donated funds and land to build Jenkins Memorial Church in Rivera Beach)

1½ - R - 21mm
7½ - R - 21mm

J. C. J. — **James Clinton Jacobs**, 1883 - 1968
Severn
(Bought farm from Thomas Beasley. Main part of farm is now part of Pioneer City Development on Reece Road. Brought pickers in from Washington, D.C. before getting them from Baltimore. Relatives believe the "R" is for Raspberries)

1R - R - 28mm - br
2R - Oc - 28mm - br
2 - Sc - 22mm - br
5R - Cl - 28mm - pr
10R - Oc - 23mm - br
20 - R - 24mm - rr
25R - R - 24mm - br - Al
50 - R - 22mm - fr
50 - R - 21mm - br

J. F. J. — **Joshua Fred Johnson**, 1883 –
Mountain Road, Pasadena
(Founded the J.F. Johnson Lumber Company. He had a small farm and the family recalls picker checks.)
30 – Oc – 25mm - br

J. H. J. — **John Henry Joyce**, late 1860's - 1905
Glen Oban
1Qt - R - 21mm - br

J. I. J. — ?
NV - R - 24mm - fr(J.I.J on both sides)

S. J. — **Summerfield Jackson**, 1863 - 1945 (Black Farmer)
Hanover/Linthicum
(Family recalls tokens)

S. T. C.of J. — (Listed under C)

T. W. J. — **Thomas Whitney Johnson**, 1882 - 1962
Farmed Marley Neck Road, Solley.
(Family recalls checks.)

W. E. J. — **William Elijah Johnson**, 1860 - 1943
Veterans Highway
(Family recalls that checks were scattered in the yard)
1.00 - Sc

W. T. J. — ?
5 - Oc - 23mm - br

Z. J. — **Zachariah Johnson**
Pasadena
(Partner with John Blanchard Neidert. See also J & N under N)
½ - R - 21mm - fr
10

"K"

K — **Joseph Koppold**
"White Oak Bottom", near Elvaton Road, Millersville.
(Family recalls tokens with a "K")

B. A. K. — **Benjamin A. Klug**
Mountain Road, Pasadena
½B - R - 22mm

C. F. H. K. — **Charles Fulton Hezron Kelly,** 1858 - 1925
Camp Meade Road, now BWI.
(Father of Lewis Kelly. Donated land for Friendship Chapel. Family recalls throwing pickers' checks down the well.)

E. K. — **Egbert Joseph Kelly,** 1863 - 1945
Dorsey Road, Harmans
(Son of HNK)
1 - R - 20mm - fr
2 - R - 21mm - br
5 - Oc - 19mm - fr
10 - Oc - 19mm - fr
15 - Oc - 23mm - br
50 - Sc - 22mm - br
100 - Sc - 25mm - br

E. J. K. — **Egbert J. Kelly,** same as EK
5 - R - 24mm - rr

H. F. K. — **Henry Fredrick Kaiss**
Old Annapolis Road, North Linthicum
(Family recalls tokens. Some had "H. Kaiss" on them)
7½ - Sc -23mm

H. N. K. — **Hezron Nehemiah Kelly,** 1818 - 1880
Dorsey Road, Harmans
(Father of EK)
1 - R - 20mm - fr
1 - R - 21mm - br(2var)
2 - R - 20mm - fr
2 - R - 21mm - br(2var)
5 - Oc - 19mm - fr
5 - R - 20mm - fr
10 - R - 20mm - fr
10 - Oc - 19mm - fr
15 - R - 24mm - rr
15 - R - 23mm - fr(2var)
15 - Oc
50 - Sc - 21mm - br
100 - Sc - 25mm - br
1000 - R - 29mm

H. S. K. — **Hiram S. Kelly,** 1859 - 1932
McPherson on Dorsey Road.
1 - R - 19mm
2 - Sc - 24mm
5 - Oc - 25mm
10 - Sc - 25mm
15 - R - 23mm
100 - R - 30mm
1000 - Oc - 28mm

J. K. **Joseph Stanislaus Kostkowski**, 1872 - 1940
Dicus Mill Road, Millersville
(Polish immigrant came to Sparrows Point at age 16. Worked at fertilizer plant at Curtis Bay. Daughter remembers tokens with J.K., values of 1, 5, 10, 20, and 30; most Round, some Scalloped, all brass purchased with initials stamped on them. They were used for strawberries, peas and beans)
5 - R - 21mm - br

J. H. K. **J. H. King**
Annapolis Junction
1 - R - 20mm - fr

L. K. **Lewis Kelly**, 1886 - 1975
Corner of Telegraph Road and Route 174
(Family recalls tokens. Son of CFH Kelly. Lewis Kelly also used checks belonging to his father-in-law, John Stewart.)

L. C. K. **Lewis Charles Knopp**
Old Oak Lane, Severn
(Bought farm from Thomas J. Gardner. Son recalls tokens)

T. R. K. ?
1 - R - 22mm
5 - R - 23mm
10 - Sc - 24mm
15 - Sc - 24mm
20 - R - 25mm
100 - Oc - 27mm
200 - Cl - 29mm

W. D. K. **Walter D. Klingelhoefer**
Mountain Road, Pasadena
(First farm near ice rink on Rt. 97. Bought the farm of Dr. Billingslea where Route 100 now intersects Mountain Road. Bought checks in 1920 and used them until 1948. All are holed as they were strung on a small chain.)
1 - R - 24mm - br(2var)
1 - R - 26mm - pr - Al
5 - R – 26mm - pr
10 - R – 26mm - pr

"L"

A. L. **Abner Linthicum** 1842 - ?
Maple Road, Linthicum
(Reported to have used checks)

A. S. L. ?
1 - R - 22mm - br

C. C. L. — **Charles C. Leffet**, died 1960
Annapolis Road (Route 175), Odenton
1½ - Oc - 19mm - br
15 - Sc - 22mm - br
30 - Oc - 22mm - br
90 - R - 23mm - fr

F. L. — **Frederick Link**
Near Beltway restaurant
½Bus - R - 24mm - fr
2 - Sc - 22mm - br
5 - Sc - 22mm - br
10Bus - R - 24mm - fr
50 - Sc - 22mm - br

J. Libaros — **J. Libaros**
2½ - R - 28mm - br (2½ on rev.)

J. M. L. — **J. M. Lowman**
Odenton
1 - Cl - 28mm - pr
5 - Sc - 28mm - pr
25 - R - 21mm - br
50 - Sc - 22mm - br
100 - Oc - 23mm - br
500 - Sw - 19x28mm - pr

M. J. L. — **M. J. Lowman** (?)
10 - R - 20mm - fr

S. L. — **Sweetser Linthicum, Sr.**, 1824 - 1905
Linthicum
(Checks also used by his son, Sweetser Linthicum, Jr.)

S.L. Jr. — **Sweetser Linthicum Jr.**, 1862 - ?
15 - R - 22mm
(Found in field by metal detector)

W. L. — **Wesley Linthicum**, 1828 - 1895
Mountain Road
(Tokens were used by son, W.L.Jr.)

W. J. L. — **William James Lockerman**
Jenkins Road, Rivera Beach
(Served in the Union Army during the Civil War. Farm sold to James A. Jenkins. Granddaughter recalls many tokens)
1½ - Sc - 22mm - br

"M"

McK — **George W. McKnew**, ? - 1917
Rt. 175 & McKnew Road, Gambrills
1 - R - 21mm - br
50 - Sc - 22mm - br
150 - Oc - 23mm - br

A. M. — **Abraham Myers**, 1867 - 1941
Quarterfield Road, Severn
(Grandson recalls tokens of 1, 5, 10, 20, 50)
1 - Sc - 25mm - br
10 - Sc - 25mm - br
50 - Sc - 25mm - br

A. G. M. — **Abraham G. Myers**
New Cut Road
½pk - Oc - 23mm - br
1 pk - Oc - 23mm - br
1 BUS - Oc - 23mm - br
5 BUS - Oc - 23mm - br
10 BUS - Oc - 23mm - br

Also used the following Aluminum checks:
½pk - Sc - 23mm - br – Al
1pk - Sc - 23mm - br - Al
1BUS - Sc - 23mm - br - Al
5BUS - Sc - 23mm – br - Al
10BUS - Sc - 23mm – br - Al

B. F. M. — **Benjamin Franklin Mewshaw**, 1880 - 1928
Brooklyn
(Son of Franklin Mewshaw)
½ - Oc - 22mm - br
1 - R - 20mm - fr
1¼ - R - 20mm - fr
2 - R - 20mm - br
8 - Sc - 22mm - br
10 - R - 20mm - br
20 - R - 20mm - br
50 - R - 24mm - rr
100 - Sc - 25mm - br

F. C. M. — ?
50 - Oc - 26mm

F. M.
F. MEWSHAW

Franklin Mewshaw, 1845 - 1903
Quarterfield Road, Brooklyn
(æ stands for Anderson-Erland hardware store in Baltimore where many checks were purchased)

NV- æ - R - 23mm - fr
½ - Oc - 19mm - fr
1 - R - 21mm - br
1¼ - R - 21mm
8 - R - 20mm - fr(2var)
8 - R - 21mm - fr

25 - R - 23mm - fr
50 - R - 23mm - fr
100 - R - 23mm - fr
100 - SC - 22mm - br

G. Mc or
G. McP

George McPherson
Dorsey Road, near the old WB & A RR
(Believed to have used checks.)

G. M.

?
(Check found in a farm field in Severn)
1 - Sc - 22mm - br

G. G. M.

?
NV - R - 19mm - pr

G. M. M.

George Montgomery Murray, 1843 - 1906
1390 Odenton Road, Odenton
(George Murray started a canning house in Odenton which was later operated by his sons, the Murray Brothers. The building was torn down in 1928 when Murray D. O'Malley started a coal business on the site. The Murray Brothers also built The Murray Hotel around 1906 which burned down in 1923. Used brass tokens and large tin tokens the size of a half dollar.)
½P - Oc - 20mm - br

H. M.

Henry J. Myers(?)
Dundee, Glen Burnie
(Henry Myers won the first Model T raffled off at the Glen Burnie Carnival.)

1 - R - 20mm - fr
5 - R - 20mm - fr
10 - R - 20mm - fr
25 - R - 20mm - fr

H. H. M.

?
½P - Oc (P above ½)
5P - R (P above 5)
5 - R

H. J. M.

Henry Jackson Myers, 1857 - 1916
Dundee, Glen Burnie
1 - Oc - 25mm - br
2 - Oc - 25mm - br
3 - Oc - 25mm - br
5 - Oc - 25mm - br
10 - Oc - 25mm - br
25 - Oc - 25mm - br
50 - Oc - 25mm - br
100 - Oc - 25mm - br

J. M.

John H. Meek, 1879 - 1936
Ft. Smallwood Road, Jacobsville
1B - R
1B - Oc
5 - Oc
5 - R - 21mm - br
5 - Sc
5B - Oc - 23mm - br
7½ - Sc(holed)
10 - Sc
10 - R - 21mm - br
10B - Oc - 23mm - br
20 - R - 21mm - br
20B - Sw - 24x28 - pr
20B - Oc - 23mm - br
50 - Sc - 27mm - pr
100 – R – 21mm - br
100 - Oc - 26mm - br
1.00 - Oc
300 - R - 21mm - br
3.00 - Oc - 25mm - br

J. S. M.

?
5 - R - 24mm - fr

L. M.

Lewis Myers
1 - Sc
2 - Sc

R. H. M.

Richard Henry Maynard, 1865 - 1919
Odenton
(Mr. Maynard also owned a canning house in Woodwardville. Family recalls tokens)

W. J. M.

?
NV - R - 29mm - fr(rev."5")

W. R. M.

William R. Mewshaw
Quarterfield Road, Glen Burnie
(Believed to have used tokens)

"N"

N ?
1BKT - R - 21mm - pr
1BKT - R - 21mm - fr(die struck)

C. A. N. **Charles Adam Neidert**, 1878 - 1954
337 Obrecht Road, Millersville
(Originated the Knight green-meat cantaloupe.
Checks reported of values: 1, 5, 10, 15, 100, 500)

J. E. Nunn **J. E. Nunn**
Severn
(reported to have used tokens)

J & N **Zechariah Johnson & John Blanchard Neidert**
Pasadena
NV - Oc - 19mm
1 - Oc - 19mm
5 - R - 22mm - fr(3var)
10 - Sh - 23x25mm - pr
25 - R - 24mm - pr
50 - Cl - 28mm - pr
100 - Sc - 25mm - br
100 - Sc - 28mm

R. G. N. ?
50 - Sc - 27mm - br

R. M. N. ?
5 - R - 24mm - br("G" counterstamped over M &"G" on rev)

"O"

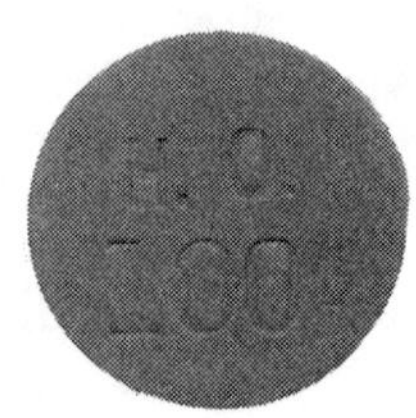

H.O. **Harry Owens**, 1891 - 1969
Smith Station, Elmhurst
(Peter & Anna Browskie served as "row bosses"
for H.O. for 20 years)
1 - R - 24mm - pr - rF
5 - R - 24mm - pr - rF
10 - R - 24mm - pr - rF
20 - R - 24mm - pr - rF
50 - R - 24mm - pr - rF
100 - R - 24mm - pr - rF

J.O. **Joshua Owens**

Race Road, Hanover

(Some checks have "H.O." counterstamped on the rev. for Harry Owens. Checks were also used by W. H. Clark. There are two distinct sets of H. O. checks. The tokens listed below on the left are confirmed to be from Joshua Owens, but the tokens on the right may have been issued by Jonas Oliver, a black farmer, who farmed on Donaldson Ave., Severn)

$1 - R - 23mm - fr	1 - Oc - 19mm - br
2 - R - 20mm - fr(2var)	2 - Sc - 22mm - br
5 - R - 23mm - br	5 - R - 21mm - fr
6 - R - 20mm - fr(2var)	10 - Oc - 22mm - br
10 - R - 23mm - br	15 - R - 23mm - fr
50 - R - 23mm - fr	20 - Sc - 25mm - br
1.00 - R - 24mm - fr	50 - Sc - 28mm - br
	100 - Oc - 26mm - br
	200 - R - 28mm - br

R. E. O. **Robert Edwin Owens**, 1859 - 1944

Ridge Road, Hanover

(red and black fiber tokens used in 1889)

1 - R - 23mm - pr - rF	10 - R - 24mm - pr - rF
1 - R - 24mm - pr - rF	10 - R - 23mm - pr - bF
5 - R - 23mm - pr - rF	10 - R - 29mm - pr - bF
5 - R - 29mm - pr - rF	20 - R - 24mm - pr - rF

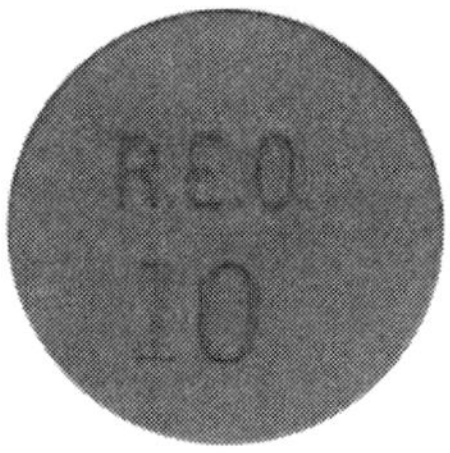

T. O. **Theodore Osborne**

Ft. Smallwood Road, Chesterfield area, Pasadena

1½ - R (2var)	15 - Sc
1½ - Oc - 19mm - fr	15 - R - 23mm - br(2var)
3 - R - 20mm - fr	30 - R - 23mm - br(2var)
4 - R - 23mm - fr(4 0n rev)	75 - Sc - 25mm - br
7½ - R - 20mm - fr(2var)	100 - R - 23mm - fr
7½ - Oc (2var)	100 - Sc
10 - Sc - 22mm - br	
10 - R	

"P"

P

?
NV - R

P. Jr.

Walter James Pumphrey, Jr., 1870 - 1907
Veterans Highway, Millersville
(Checks were also used by his son, Earl James Pumphrey, 1896 - 1960.)
1 - R - 21mm - fr
50 - R - 23mm - fr
100 - Sc - 26mm - br

A. P.

Addison Pumphrey
Severn Run
(Addison "Ad" Pumphrey, brother of Benjamin F. Pumphrey, Sr. owned a tomato cannery at Severn Run. Brand name, "Pride of Benfield." He may have used tokens in his cannery.)

A. C. P.

Aloysis Clinton Pumphrey
Ridge Road, Hanover
(Forrest Huffmaster found 32 checks in a bag in an old barn on Ridge Road in 1953. He donated 19 tokens to the AACo. Hist. Soc.)
1 - R - 20mm - fr
3 - R - 20mm - fr
5 - R - 20mm - fr
10 - R - 20mm - fr

B. F. P.

Benjamin Franklin Pumphrey 1854 - 1927
Millersville
(Benjamin F. Pumphrey was a carpenter by trade and left the farm for his sons, Benjamin F. Jr., Nathanial Asbury, and Linton Abraham, to operate. The road to Ben's fields became known as Benfield Road. BP is for Big Peas, LP is for Little Peas.)
NV - R - 21mm - br
¼L.P.-R-21mm - br
¼B.P.-R-21mm - br
½L.P.-R-21mm - br
½B.P.-R-21MM - br
1 - R - 21mm - br
1 - R - 20mm - fr(2var)
2 - Oc - 19mm - fr
2B.P.-Sc-22mm - br
2L.P.-Sc-22mm - br
4 - R - 20mm - fr
5L.P.-Oc- 23mm - br
5B.P.-Oc- 23mm - br
6 - R - 21mm - br
10 - Sc - 22mm - br
10B.P.-Oc-25mm - br
10L.P.-Oc-25mm - br
25 - Sc - 22mm - br
100 - Oc - 23mm - br

C. P. **Charles D. Pumphrey**
NV - Sc - br

C. A. P. **Charles Abraham Pumphrey**, 1891 - ?
Mountain Road
(Family recalls tokens)

C. B. P. **Charles Benjamin Pumphrey**, 1827 - ?
Marley, Sun Valley development
(Grandfather of Ora P. Smith)
NV - Sc - 25mm - br(2var)
1 - Oc - 19mm - fr
1 - R - 20mm - br
10 - R - 24mm – rr(2var)
10 - Sc - 25mm - br
10 - Sc - 24mm - br(2var)

C. G. P. **Charles Greenberry Pumphrey**
Veterans Highway
(Called "One Eye Charles" as he lost an eye in a hunting accident at the age of 16)
15 - R
20 - R

E. L. P. ?
5 -

F. P. **Frank Pumphrey**
(Notice in *The Evening Capital*, May 21, 1896)
500 - R

H. P. ?
15 - Oc - 26mm - br

J. P. **John Pumphrey** (?)
1 – R – 20mm - fr
25 - R - 20mm - fr
500 - R - 20mm - fr

J. D. P. **Jefferson Davis Phelps**, 1860 - 1940
Aviation Blvd. & Dorsey Road, Harmans. Now within BWI.
15 - R - 24mm
1000 - R - 27mm

J. E. P. **James E. Phelps,** 1858 - 1936
Between Camp Meade Rd. and Dorsey Road at Kelly Station on the B&A Railroad. Now within BWI. (Checks were donated for scrap metal to support the war effort. Only one check known to survive.)

J. R. P. **John Randolph Parker**, 1846 - 1912
Crain Highway, Glen Burnie
(Operated farm and the 3rd District Store which opened around 1890. The store was located on the East end of his home.)
NV - R - 34mm - pr - Fe

The following tokens were die-struck for the 3rd District Store:
5 - R - 18mm - fr (rev eagle)
10 - R - 19mm - fr (rev Liberty Head)
25 - R - 24mm - fr (rev Liberty Head)

K. P. **Knollie Grafton Pumphrey**
"Pumphrey's Search", 162 Olen Drive, Ferndale
5 - Sc - 23mm - br
20 - R - 21mm - fr
25 - Sc - 26mm - br
500 - Oc - 23mm - br

L. B. P. ?
NV - R - 24mm - fr

N. P. **Nelson S. Phelps**, 1894 - ?
Wellhams
(Family recalls tokens)

N. A. P. **Nathaniel Asbury Pumphrey**, 1887 - 1962
8300 block of Veterans Highway, Millersville
(Son of B.F.P. 39 checks were donated to the AACHS.)
1 - Oc - 18mm - pr
2 - R - 19mm - pr
5 - Sc - 22mm - br
10 - Oc - 23mm - br
25 - Sc - 28mm - br
100 - Sc - 28mm - br

N. C. P. — **Nathan C. Pumphrey**, 1821 - 1891
Furnace Branch Road, Pleasantville
(Checks were also used by Carl Shipley. Dan Patterson dug up 254 NCP checks in 1983. Most were donated to the AACHS)
1½ - R - 20mm - fr
7½ - Sc - 22mm - br
15 - R - 23mm - fr
30 - R - 22mm - br
75 - Sc - 25mm - br

O. C. P. — **Oden C. Pumphrey**
Second District
(Ads in Evening Capital, 1896-1900, give notice that Oden C. Pumphrey will only redeem his checks from certain persons.)

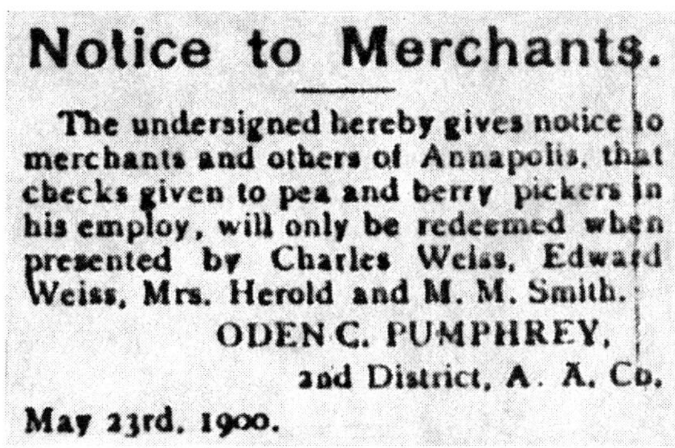

Notice to Merchants.

The undersigned hereby gives notice to merchants and others of Annapolis, that checks given to pea and berry pickers in his employ, will only be redeemed when presented by Charles Weiss, Edward Weiss, Mrs. Herold and M. M. Smith.

ODEN C. PUMPHREY,
2nd District, A. A. Co.

May 23rd, 1900.

O.S. P. — **Osborne Stallings Pumphrey**
Farmed where Marley Station Mall is now located.

R. D. P. — **Rufus Dallas Phelps**, 1845 - 1918
Ritchie Highway(DMV), Glen Burnie
½ - R - 23mm - fr
20 - R - 23mm - br(rev"RNP 20")
100 - R - 23mm - fr(rev"RNP 100")

R. H. P. — **Richard H. Phelps**
Mountain Road, Pasadena
(Farm purchased by the Schramms. Two RHP tokens were found in the field.)
3 - R - 22mm ("4" on rev.)
10 - R

R. L. P.

Robert Lee Pumphrey, 1867 - 1934
Jumpers Hole Road, Pasadena

NV - Sc - 25mm - br
1 - R - 21mm - br(2var)
1 - Oc - 19mm - fr
10 - R - 24mm - rr(2var)
10 - R - 24mm - fr
10 - Sw - 24x28mm - pr
50 - Oc - 26mm - br
100 - R - 29mm - br

R. N. P.

Roland Nathaniel Phelps, 1873 - 1953
Ritchie Highway, Glen Burnie

½ - R- 23mm - fr
1 - R - 22mm - br - Al(2var)
5 - Sc - 22mm - br - Al(2var)
5½B - Sc - 25mm - br(2var)
10 - R - 21mm - br(rev"10")
15 - R - 21mm - br(rev"15")
20 - R - 24mm - rr(rev"RNP 20")
25 - Oc - 23mm - br - Al
100 - R - 23mm - fr(rev"RDP 100")

S. T. P.

Sedwick Thomas Pumphrey, 1864 - 1929
Old Annapolis Road, Ferndale

1 - R - 20mm - fr
5 - Sc - 22mm - br
7½ - Sc - 22mm - br
10 - R - 21mm - br
20 - Sw - 24x28mm - pr
75 - Sc - 22mm - br
100 - Oc - 22mm - br
500 - R - 24mm - br
600 - Sc - 22mm - br

T. P.

Thomas Pumphrey, 1824 - 1909
Furnace Branch Road, Pumphrey's Station, Ferndale
(Ora Smith's great uncle.)

1 - R - 20mm
1s - Sc - 22mm - br
1sp - Sw - 24x28mm - pr
2s - Sc- 22mm - br
3 - Oc - 23mm - br
3 - R - 22mm
3s - Sc - 22mm - br
3sB - Sw - 24x28mm - pr
4sp - Sw - 24x28mm - pr
5sB - Sw - 24x28mm - pr
5 - Oc - 19mm
10 - R - 22mm
10 - Oc - 23mm - br
10s - Sc - 22mm - br
10sp - R - 21mm - fr
15s - Sc - 22mm - br
15sp - Sw -24x28mm - pr
20 - R - 21mm
20 - Oc - 23mm - br
20s - Sc - 22mm - br
20sb - Sw - 24x28mm - pr
25 - Oc - 23mm - br
25s - Sc - 22mm - br
50 - Sc - 22mm
50s - Sc - 22mm - br
100 - Oc - 23mm - br
100sp - Sw - 24x28mm - pr

T. C. P. **Thales Cornelius Pumphrey**, 1862 - 1942
"Pumphrey's Search," Pumphrey's Station, Ferndale
(Family has tokens. Ora Smith's father.
Farmed next to Cromwell Shopping Center)
2½sp – R – 22mm – br
5BP – Oc – 23mm – br
50BP – Oc – 23mm – br

T. O. P. **Thomas Octavius Pumphrey**, 1841 - 1909
Benfield Road, Millersville
(Thomas Pumphrey never farmed. The farm was operated by his sons. Checks were also used by the sons, James S. Pumphrey, John, and Walter James. They used German pickers from Curtis Bay. Brother of Benjamin Franklin Pumphrey)
1 - Oc - 19mm - fr
3 - Oc - 19mm - fr
5 - R - 21mm - br
10 - R- 21mm - br
15 - Oc - 22mm - br
50 - Oc - 22mm - br
100 - Oc - 25mm - br

W. P. ?
5 R - 20mm - fr

W. J. P. **Walter James Pumphrey, Sr.**, 1857 - 1890s
Veterans Highway, Millersville
(Grandson, Earl Pumphrey, also used his checks)
100 - Sc - 26mm - br

W. J. P. **William John Pumphrey**, 1876 - 1947
500 block Veterans Highway
(Used a red box to house his checks)
NV - R - 21mm
1 - R - 21mm
3 - R - 21mm
5 - Sc - 23mm
10 - Sc - 23mm
20 - Oc - 23mm - br
50 - Oc - 23mm - br
100 - Cl - 28mm
100 - Sc - 26mm
500 - Cl - 28mm

W. M. P. **William Marman Pumphrey**, 1864 - 1918
8th Ave., Glen Burnie
(Family recalls checks)

W. W. P. **William Wallace Pumphrey**, 1862 - ?
(Red Fiber checks owned by family)

1 - R - 21mm - pr - rF	10 - Cl - 29mm - pr - rF
3 - Sc - 22mm - pr - rF	50 - Oc - 26mm - pr - rF
5 - Oc - 23mm - pr - rF	100 - Sc - 29mm - pr - rF

"R"

R. **Roberts** (?)
NV - R - 20mm - fr
50 - R
100 - R -(rev"100" 2var)
100 - Sc

B. T. R. **Benjamin Thomas Ray**, 1846 - 1921
Harmans

1 - R - 23mm - fr(2var)	8 - R - 23mm - fr
1 - R - 24mm - rr	8 - R - 24mm - rr
2 - R - 23mm - fr(2var)	10 - R - 23mm - fr(2var)
2 - R - 24mm - rr	10 - R - 24mm - rr
4 - R - 23mm - fr	15 - Oc - 22mm - br
4 - R - 24mm - rr	50 - Oc - 26mm - br(2var)
5 - Sc - 22mm - br(2var)	100 - Sc - 28mm - br

E. W. W. R. **Ernest W. W. Ruth**
(Reported to have used tokens)

F. S. R. **Frank S. Revell**
Marley
(Farm was still operating in 1994)
NV - Sc - 26mm
1½ - Oc - 20mm

J. R. **John Richard Rice**, 1856 - 1944
Waterbury, Crownsville

1 - Sc - 22mm	5 - R - 25mm - pr - rF
5 - Sc - 22mm	10 - R - 25mm - pr - rF
10 - Oc - 23mm	25 - R - 25mm - pr - rF
	50 - R - 25mm - pr - rF

J. B. R. **Joseph Benson Ray**, 1881 - 1962
Severn

1 - R - 21mm - br
2 - R - 21mm - br
5 - Oc - 25mm - br
8 - Oc - 25mm - br
10 - Oc - 25mm - br
15 - Sc - 29mm - br
50 - R - 29mm - br
100 - R - 35mm - rr

J. H. R. **John Henry Robinson**, 1844 - ? (Black Farmer)
River Road, North Linthicum
(When freed in 1864, his owner gave him land. He planted truck produce and bamboo for brooms. See text for more information on John Henry Robinson.)

1 - Oc - 19mm - pr
2 - Oc - 19mm - pr
10 - Oc - 23mm - br
20 - Oc - 23mm - br
25 - Sc - 28mm - pr
25 - Sc - 31mm - pr
100 - Sc - 29mm - br

J. S. R. ?
(Some have WTR on reverse)
NV - R - 21mm - br(2var)
NV - R - 23mm - fr
NV - R - 24mm - fr
15 - Sc - 23mm - br

J. W. R. **J. W. Ray**
Severn
1 - Oc

R. R. **Romeo Redmiles**
Telegraph Road, Severn
(Family reported that RR used homemade checks of different sizes and shapes. No initials, just values,"about the size of a postage stamp." None are known to exist. Mr. Redmiles also sold milk from an old truck)

W. R.
W. Remmey **William Remmey**, 1856 - 1927
Brooklyn
(Most checks are counter stamped "F.M."as they were also used by Franklin Mewshaw. The 1-R, ½{19mm} and 8-R tokens are marked, "W. REMMEY". 271 checks were donated to the AACHS in 1990)

½ - Oc - 19mm - fr(FMrev)
½ - Oc - 23mm - br(FMrev)
1 - R - 21mm - br(FMobv)
1 - Oc - 23mm - br(FMobv)
5 - R - 23mm – fr(noFM)
8 - R - 20mm - br (FMobv)
8 - Oc - 23mm - br (FMobv)
10 - Sc - 25mm – br (noFM)
100-Oc - 23mm - br (FMrev)

W.T.R. ?
NV - R - 24mm - fr(2var)
NV - Sc- 22mm - br
10 - Oc - 23mm - br
15 - Oc - 23mm - br
20 - Oc - 23mm - br

"S"

SHRIVER SHRIVER/ BROS. / FARMERS / BODKIN CREEK, MD.
(reverse) GOOD FOR / 2 / IN TRADE
Al, 20mm
(Same but values and sizes: 5 - 24mm, 10 - 25mm, 25 - 27mm, 50 - 30mm, 1.00 - 34mm)
(Shriver brothers, George, James, William, Sidney, & Charles, purchased farm in 1928; operated until 1960 when George & Charles bought a small farm in Dorchester County. Last surviving brother, Charles died in February 1997. Tokens are die struck)

A. S. **Allen Stevenson**
(One token found in a field near Glen Burnie)
10 - R

A. A. S. **Adam Alexander Shipley**, 1844 - 1915
Ridge Road, Harmans
(All tokens are aluminum and are badly corroded as they were recovered from a well where the family had disposed of them)

1 – R – 22mm – br - Al
2 – R – 22mm – br - Al
5 – R – 22mm – br - Al
10 – R – 22mm – br - Al
16 – R – 22mm – br - AL
20 – R - 22mm – br - Al
50 – R – 22mm – br - Al
100 – R – 22mm – br - Al

A. A. S. **Alfred Asbury Stinchcomb**, 1848 - 1915

St. Margarets, Bay Head on west side of Little Magothy River. (Cape St. Clair) (335 tokens were donated to the AACHS. Two distinct series of tokens exist. One series conforms with standard procedures. The other series has the values on the reverse and are all octagonal.)

NV - Oc - 29mm – br
1 - R - 20mm – br
1 - Sc - 22mm – br
1½- R - 24mm - rr
2 - Sc - 22mm - br
3 - R - 24mm - rr
4 - Oc - 22mm - br
5 - Sc - 22mm - br

10 - R - 21mm – br
10 - Sc - 22mm – br
10 - Sc - 25mm - br(2var)
15 - R - 21mm - br
20 - R - 21mm - br
20 - Sc - 22mm - br
30 - R - 24mm - rr
50 - Sc - 22mm - br
75 - R - 29mm - br
500 - R - 28mm - br(AAS in script)

(series two - with values on reverse)

¼Bu - Oc - 25mm - pr
½Bu - Oc - 25mm - pr
1Bu - Oc - 25mm - pr
2½Bu - Oc - 25mm - pr

5Bu - Oc - 29mm - br
10Bu - Oc - 29mm - br
20Bu - Oc - 29mm - br

A. C. S. **August Carl Schmidt**

Pinehurst, Pasadena
NV - R - 20mm - pr - rF
NV - R - 24mm - rr

B. H. S. **Basil Hamilton Smith**, 1884 - 1960

Telegraph Road, Severn
(John & Ora Smith's farm)
1 - R - 21mm - pr
2 - R - 24mm - pr
200 - Sc - 28mm - br

C. H. S. **Carl Hawkins Shipley**, 1885 - ?

Furnace Branch Road, Glen Burnie
(Family recalls tokens)

C. L. S. **Charles Luther Solley**, 1860 - 1921

Solley Road, Glen Burnie
2 - Sc - 22mm - br
10 - Sc - 22mm - br
20 - SC - 22mm - br

C.W.S. **Charles Winfield Stinchcomb**
(Son, Rufus Oliver Stinchcomb, also used his father's checks)

C.V.B. Smith **Charles Van Buren Smith**
Camp Meade Road now Telegraph Road, Severn
(127 checks donated to AACoHS)
1 – R – 21mm - br
2 – R – 21mm - fr

D. St **Daniel Stoll**
Old Ordnance Road
(Farmhouse was torn down to build the Baltimore Beltway in 1969.
2 - R - 35mm - pr - I
4 - R - 35mm - pr - I
6 - R - 35mm - pr - I
8 - R - 35mm - pr - I
10 - R - 35mm - pr - I
16 - R - 35mm - pr - I
25 - R - 21mm - fr

E.S. **Emil Schultz**
Reece Road, Severn
(Thought to have used checks)

E. S. **William Elbridge Shipley**, 1870 - 1930
Ridge Road
(Son of Adam A. Shipley. Used both brass and aluminum tokens. The aluminum tokens were made after 1930 and were used by sons, Irvin E. and Edwin A. Shipley)

Brass Tokens:
1 - R - 21mm - br
5 - R - 21mm - br
5 - R - 22mm – pr
10 – R – 22mm – pr
10 – R – 21mm – br
20 - R - 22mm - pr
50 - R - 29mm - pr
1.00 – R – 29mm - br
1<u>oo</u> – R – 29mm - pr

Aluminum Tokens:
½ - R - 19mm - pr - Al
1 - Oc - 19mm - pr – Al
2½ - Sq - 21mm - pr - Al
5 - R - 24mm - pr - Al
10 - Oc - 21mm - pr - Al

E. A. S. ?
1BKT - R - 21mm - pr

E. L. S. — **Eleza Lee Stewart**, 1879 - 1929
Crain Highway, Glen Burnie
(Son of J. C. Stewart. Original farm was on Quarterfield Road and Rt. 3. Farm was sold to Cromwell Whittemore, son of Arthur Cromwell Whittemore. After selling farm, Eleza Stewart moved to 5th Ave., Glen Burnie)

½B - Sc - 22mm - br
½P - Oc - 23mm - br
1 - R - 21mm - br(2var)
1 - R - 20mm - fr
2 - R - 21mm - br(2var)
2 - R - 20mm - fr
3 - R - 21mm - br(2var)
3 - R - 20mm - fr
4 - R - 21mm - br
5 - R - 21mm - br(2var)
5 - R - 20mm - fr
5B - Sc - 22mm - br
5P - Oc - 23mm - br
8 - R - 21mm - br(3var)
8 - R - 20mm - fr
10 - R - 20mm - fr
10 - R - 21mm - br(2var)
10B - Sc - 22mm - br
10P - Oc - 23mm - br
15 - R - 21mm - br
25 - R - 21mm - br
30B - Sc - 22mm - br
30P - Oc - 23mm - br
100 - R - 28mm - br
200 - R - 28mm - br
300 - R - 28mm - br
500 - R - 28mm - br
1000 - R - 28mm - br
1500 - R- 35mm - pr
2000 - R - 35mm - pr

F. S. — **Frank Sewell** (?)
½P - R - 21mm - br

F. M. S. — ?
25 - R - fr

F. O. S. — **Francis Olander Spriggs**, 1872 - 1930
Arnold
(Son of Edward Spriggs)
10 - R - 23mm - fr
50 - Oc - 22mm - br
100 - Sc - 25mm - br

G. S. — ?
½B - Oc

G. B. S. — **George Bandell Stinchcomb**, 1864 - 1945
Manhattan Beach Road
(Tokens recovered from a well. Father-in-law of George A. Jenkins)
5 - Oc - 19mm - br
100 - R - 23mm - fr

G. C. S. — **George Carl Schmidt**, 1891 - 1975
Mountain Road, Pasadena

NV - R - 24mm - rr
1 - R - 20mm - fr
1 - R - 21mm - pr
1 - R - 22mm - fr
5 - Sc - 22mm - pr
10 - Oc - 20mm - pr
20 - Sw - 24x28 - pr
100 - Oc - 23mm - br

G. M. & S. — **George Macey & Sons**
Old Mill Bottom Road
(Ran a cannery - "Village Brand Tomatoes")
2½ - R - 23mm - pr

H. S. — **Heinrich Sellner**, 1886 - 1958
Sellner Road, Jessup
(Born in Germany. Known as "Henry"in the U.S. He was sponsored by Max Blob and worked for Mr. Blob until his debts were paid off. He bought the farm next to Max Blob. Also used cardboard chits; only known card is: H.SELLNER - 100 Quarts-46x32mm-light blue)

1 - R- 19mm
2 - R - 19mm
3 - R - 19mm(holed)
5 - R - 19mm
10 - R - 19mm
10 - R - 19mm(holed)
20 - R - 19mm(holed)

H. S. — **Hezekiah L. Shipley**
(Probably same as H.L.S.)
½ - Sc - 22mm - br
5 - R - (found in field)

H. B. S. — **Henry Bernard Stallings** 1848 - ?
Farmed "Green Gables"at Pinehurst on Mountain Road, just east of Ventnor Road.
(Thought to have used checks)

H. J. S. — **Henry J. Schmidt**
Clark Station Road, Severn.
NV - Sq - 23mm - pr
1 - Sc - 25mm - br
1 - Sc - 25mm - pr
25 - R - 28mm - br

H. L. S. — **Hezekiah Llewellyn Shipley**, 1847 - 1912
Dorsey Road, Harmans

1 - R - 21mm - fr
5 - Sc - 22mm
100 - Oc - 26mm
200 - R - 28mm
200 - R - 30mm

J. S. **John Smith**
(Tokens found by metal detector in field)
1 - R - br
2 - Sc - 22mm - br
10 - Sc - 25mm - br

J. S. **John Stewart**, 1847 - 1934
Reece & Telegraph Road, Severn
(Bought farm from Joshua Linthicum around 1900. Checks were later used by his son-in-law, Lewis Harman Kelly, Sr. Some tokens have a large "R" on the reverse)
1 - R - 21mm
5 - R - 20mm - fr
5 - R - 21mm - rr
5 - R - 24mm - br
10 - Sc - 22mm
25 - Oc - 24mm

J. C. S. **Joshua Cromwell Stewart**, 1821 - 1898

Crain Highway, Glen Burnie
1 - R - 20mm - fr(2var)
2 - R - 20mm - fr
5 - R - 20mm - fr(2var)
8 - R - 20mm - fr
10 - R - 20mm - fr(2var)
25 - R - 20mm - fr
25 - R - 23mm - fr
50 - R - 20mm - fr
100 - R - 20mm - fr
200 - R - 20mm - fr
300 - R - 20mm - fr
400 - R - 20mm - fr
500 - R - 20mm - fr

J. E. S. **John Edward Stoll**, 1863 - 1947

Hammonds Ferry Road, North Linthicum
1 - Oc - 22mm - br
1B - R - 20mm - fr
1B - Cl - 28mm - pr
2 - Oc - 22mm - br
3 - Oc - 22mm - br
3 - Oc - 26mm - br
5 - Oc - 22mm - br
10 - Oc - 22mm - br
10 - Oc - 25mm - br
15 - Oc - 22mm - br
15 - Oc - 25mm - br
25B - Sc - 22mm - br
50B - Oc - 22mm - br
100B - Oc - 22mm - br
100 - Oc - 22mm - br

J. F. S. — **John Franklin Shipley**, 1874 - 1951
Near McPherson Station; later bought farm at Bayside, Pasadena
(Son of Adam Alexander Shipley. Originated the Long John cantaloupe.)

NV - R - 30mm - pr(hole)
1 - Oc - 10mm - br
2 - Sc - 22mm - br
5 - R - 21mm - fr
10 - Oc - 23mm - br
15 - R - 23mm - fr
20 - Sc - 26mm - br
50 - Oc - 26mm - br
100 - R - 29mm - br
500 - Sc - 29mm - br

J. W. S. — **John Wilson Shipley**, 1865 - 1952 (son) and
John William Shipley, 1832 - 1912 (father)
Harmans

1 - R - 20mm - fr
1 - R - 21mm - br
2 - Sc - 23mm - br
3 - R - 22mm - rr
3 - R - 24mm - fr
3 - Oc - 23mm
5 - R - 24mm - rr
5 - Sc - 22mm - br
5 - Sc - 26mm
10 - R - 24mm - pr
10 - R - 23mm - fr
15 - Sc - 29mm - pr
15 - Cl - 29mm
50 - Oc - 22mm - br
100 - Sc - 25mm - br

L. SCHR — **Louis Schramm, Sr.**, 1851 - 1940
Mountain Road, Pasadena, at Catherine Avenue
(Developed the "Schramm" cantaloupe. Most L.SCHR tokens are holed as they were used later as turkey tags. Schramms raised turkeys until 1990)

½BUS - R - 24mm - fr(rev"H")
2½BUS - R - 24mm - fr(# on rev"H")
5 - R - 24mm - fr
10 - R - 24mm - fr(# on rev"H")
15 - R - 24mm - fr(# on rev"H")
"1 Box Strawberries" - R - 35mm - pr - I

L. J. S. — **Lemuel Joshua Shipley**, 1851 - 1927
Harmans, now Hanover P.O.
5p - R - 25mm

L. R. S. **Larkin Rodolphus Shipley**, 1841 - 1890
Harmans
(Tokens used 1874 - 1940. Checks used by sons: Marvin R. Shipley, Charles W. & Irvin L. Shipley. Round and Octagonal tokens are the oldest. Sw tokens purchased around 1900 to replace the 10's were called "the new checks")

¼p - R - 21mm - br
½p - R - 21mm - br
1p - R - 21mm - br
1 - R - 21mm - br(2var)
2 - R - 21mm - br(2var)
2½p - R - 21mm - br
5p - R - 21mm - br
5 - R - 21mm - br(2var)
5 - Sc - 22mm - br
10 - R - 20mm - fr
10p - R - 21mm - br

10 - Sw - 24x28mm -pr
15 - Oc - 23mm - br
15 - Oc - 22mm - br
20p - R - 21mm - br
50 - R - 24mm - fr
100 - Sc - 25mm - br(2var)
500 - R - 21mm - br
500 - Oc - 25mm - br

M. S. **Michael Sroka**
Ridge Road, Hanover

NV - R - 20mm - br
1B - R - 22mm - fr
1P - R - 22mm - fr
1B - R - 21mm - br
1BKT - Oc - 25mm - pr
1 - Sc - 22mm - pr(2var)
5 - Sc - 22mm - br
10p - R - 22mm - fr

10 - Sc - 23mm - br(2var)
15 - Sc - 22mm - br
20B - R - 22mm - fr
50 - Sc - 22mm - pr
50 - R - 24mm - rr
100B - Sc - 25mm - br
100- Sc - 22mm - br(2var)

M. F. S. **Mary Frances Spriggs**, 1848 - 1934
Arnold
(Farm started on 34 acres by husband, Edward V. Spriggs (1850-1884). Son Franklin O. Spriggs (1872-1930) also farmed. One of the few checks with a woman's initials)

1 - R - 24mm - fr
2 - R - 24mm - rr
5 - R - 24mm - fr
12 - R - 24mm - fr
25 - R - 24mm - fr
50 - R - 24mm - fr

N. S. **Nicholas Simms** , (Black Farmer)
Race Road, Hanover
5 - Oc - 22mm - pr - Fr

N. J. S. **Nicholas Joshua Stinchcomb**, 1876 - ?
(Farmed with brother, Rezin Stinchcomb. Used picker checks).

P. L. S. ?
1 - R - 24mm - rr
1M - S
1S - Sc
10 - R

R. A. S. ?
20 L.P. - Oc - 29mm
100L.P. - R - 29mm

R. L. S. **Richard Luther Shipley Sr.**, 1849 - 1923
Shipley
(Tokens were also used by his son, RLS, Jr.1880 - 1966)
1 - R - 21mm - br
1 - R - 20mm - br
5 - R - 20mm - fr
10 - R - 20mm - br

15 - R - 20mm - fr
50 - R - 23mm - fr
50 - Oc - 25mm - br
100 - R - 23mm - fr

R. O. S. **Roderick Octavius Shipley**, 1848 - 1922
Ridge Road, Harmans
(Checks die struck, values incused)
NV - R - 24mm - fr(rev "10" or "1")
1 – R – 23mm - fr
1 - R - 24mm - fr(2var)
5 - R - 24mm - fr(2var)
10 - R - 24mm - fr(3var)
20 - R - 24mm - fr(2var)
100 - R - 24mm - fr(rev "1903")

R. O. S. **Rufus Oliver Stinchcomb**, 1896 - 1987
Veterans Highway near Old Mill Rd., Millersville
(Son of Charles W. Stinchcomb, used his father's checks)
10 (on rev.) - R - 24mm
100 - R - 23mm - fr

T. H. S. — **Theodore H. Stinchcomb**

Mountain Road, Chesterfield, Pasadena
(Son of Alfred Asbury Stinchcomb. Tokens are homemade from copper, aluminum and bronze. All are oblong, 41x23mm, and have both values and a dot code or Roman numerals on the reverse. They all have a plain rim)

1 - Ob - Cu - rev "."
2 - Ob - Cu - rev ".."
3 - Ob - Cu - rev "..."
4 - Ob - Al - rev "...."
5 - Ob - Al - rev "V"
10 - Ob - Al - rev "X"
15 - Ob - Br - rev "VX"
20 - Ob - Br - rev "XX"

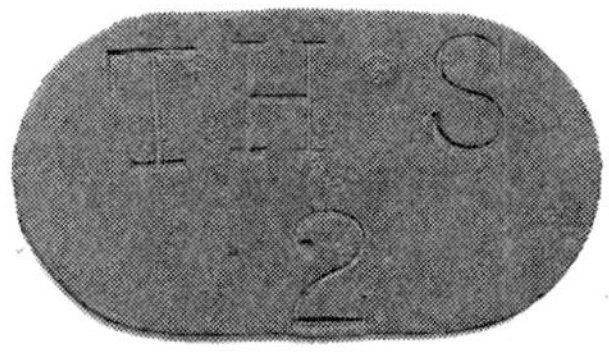

T. S. S. — **Thomas S. Simms**(?)

(A black farmer)

2 - R - 20mm
5 - R - 20mm
10 - R - 21mm
25 - R - 21mm

W. S. — **William Scott**(?)

25 - R - 28mm - br

W. A. S. — **William Abner Shipley**, 1845 - 1907

"Bachelor's Hope," Shipley
1 - R - 20mm - fr
5 - Sc

W. E. S. — **William Elbridge Shipley**

(Listed under "E. S.", as tokens were stamped without the "W")

W. J. S. — **William John Schmidt**, 1889 - 1962

Schmidts Lane off Mountain Road, Pasadena
(Son of ACS; farm was purchased in 1920. Red fiber tokens)

NV - R - 24mm - pr - rF(2var)
10 - Cl - 28mm - pr - rF(3var)
20 - Cl - 28mm - pr - rF
100 - R - 32mm - pr - rF(3mmthick)

W. J. S. — **W. J. Simms**(?)

Dorsey Road, Shipley Corner area, Harmans
(A black farmer)

NV - R
10 - Cl
20 - Cl
100 - R

"T"

T.

?
NV

B.F.T.

Benjamin F. Thomas, 1872 - 1964
Hammonds Lane, Ferndale
"B" - Sc - 22mm - br
1 - R - 23mm - fr
1 - R - 21mm - br
5 - R - 23mm - fr
10 - R - 23mm - fr
15 - R - 23mm - fr

C.R.T.

Carroll Rossiter Thomas, 1889 - 1949
Sleepy Hollow Road, Gambrills (used Odenton P.O.)
(Also farmed on Hog Farm Road - "Old Burkman Place" on New Cut & Gambrills Roads)
1 - R - 24mm - rr
2 - R - 24mm - rr
5 - R - 24mm - rr
10 - R - 24mm - rr
20 - R - 24mm - rr
50 - R - 24mm - rr

D. St

Daniel Stoll
(Tokens listed under "S")

J.F.T.

James Frank Turner
Gambrills
1 - R - 21mm
5 - Sc - 23mm
10 - Oc - 23mm
25 - Sc - 25mm
50 - R - 29mm

L.C.T.

Louis Carl Tepper, 1871 - 1942
Telegraph Road, Severn
(Family claims checks were used for all crops and were good at Rose's Store. Some checks were stamped "L.C.T. & Son" and some were stamped "Tepper")

L. J. T. — **John Linton Tubbs**, 1861 - 1930

Dorsey Road, "Wilderness Resurveyed, " Harmans

(Mr. Tubbs' tokens were erroneously stamped "L.J.T." as he was called by his middle name of "Linton".)

1 - R - 21mm - fr
2 - Oc - 23mm - br
3 - R - 21mm - fr
4 - R - 21mm - br
5 - R - 21mm - br
10 - Sc - 22mm - br
15 - R - 21mm - br
25 - Oc - 23mm - br
50 - Sc - 22mm - br
100 - R - 23mm - fr

O.T. — **O. T. Tepper** (?)

1 - Sc
2 - Sc
15p - R
20 - Sc

S. S. T. — **Sewell Summerfield Tracey**

Marley Creek

1 - R
2 - Oc
5 - R

T. J. T. — ?

2p - Oc - 23mm - br

Z.T. — **Zachariah Turner**

Crofton

(History of Crofton by Joseph L. Browne states Z.T. used tokens. He purchased the land where Crofton Mews is now located and used 25-30 black workers from Conaways to pick his crop).

"U"

J. W. U. **John Wesley Upton**, 1854 - 1910
New Cut Road, Severn
(Tokens were also used by his son, Edward Lee Upton, 1878 - 1966. Grandfather of Melvin Upton who still farms on New Cut Road.)
1 - Sc - 23mm - br
1bp - R - 21mm - br(2var)
1Lp - Oc - 23mm - br
2 - Sc - 22mm - br
2 - Sc - 23mm - br
3 - Sc - 22mm - br
5 - Sc - 22mm - br
10 - Sc - 23mm - br
50 - Sc - 23mm - br
100 - Sc - 22mm - br

T. B. U. **Thomas Bernard,** 1883 - 1950
New Cut & Upton Roads, Severn
(Used picker checks)

"V"

V. **William Harrison VanMeter**, 1911 - 1994
Crain Highway, Glen Burnie
(His father bought farm in 1907 for $300; came from West Virginia. Checks had a big "V" on them with denominations)

J.V. ?
1.00 - R – 19mm - pr

"W"

A. C. W.

Arthur Cromwell Whittemore, 1865 - 1943
Brandon Shores, Greenland Beach
(Father of Arthur Cromwell Whittermore, 1900 - 1947. Two series of tokens exist; one made of red fiber and the other of bronze. "H. O." on some tokens may stand for Harry Owens, who may also have used those tokens)

Red Fiber:

NV – R - rF
1½ - Oc - 22mm – pr - rF
1½ - R - 23mm – pr - rF
2 - Sc - 23mm – pr - rF
7½ - R - 21mm – pr - rF
10 - R - 25mm – pr - rF
15 - Sc - 29mm – pr - rF
30 – R – 29mm – pr - rF
30 - Cl - 29mm – pr - rF
50 - R - 29mm – pr - rF
100 – Sc – 29mm – pr - rF

Bronze:

NV - R - 20mm - fr
1 - R - 23mm - fr
1½ - R - 23mm - fr(2var)
10 - R - 20mm - fr(2var)
10 - R - 23mm - fr(2var"HO"rev)
15 - Sc - 22mm - br
30 - R - 23mm - fr(Some have "HO" on rev.)
¢3.00 - Oc - 26mm - br("HO"rev)
100 - Sc - 25mm - br("HO"rev)
1200 - Oc - 26mm - br

C. W.

Cromwell Whittemore, 1900 - 1947
Glen Burnie Park area
(Son of ACW. Bought farm from Eleza Stewart. Also used both bronze and red fiber checks)

Red Fiber:

1 - R - 21mm – pr - rF
2 - R - 21mm – pr - rF
2 - Oc - 23mm – pr - rF
5 - Sc - 23mm – pr - rF
10 - Cl - 29mm – pr - rF
10 - R - 21mm – pr - rF

Bronze:

1 - R - 22mm
1BUS - R - 22mm - pr
2BUS - Sc - 22mm - pr
5BUS - Oc - 23mm - br
15BUS - Cl - 28mm - pr
30BUS - Sc - 25mm - br
50BUS - Oc - 23mm - br
100BUS - R - 24mm - rr

C. R. W. **Charles Ritchie Winterson**
Hanover
1 – R – 20mm - fr

E. W. **Ernest Wesley**
Dorsey Road
(Reported to have used checks)

E. W. **Elijah Wesley Wood**, 1849 - 1925
Elvaton Road, Millersville
1 - Sc - 22mm - br
5 - Sc - 22mm - br
10 - Sc - 22mm - br(2var)
25 - Sc - 22mm - br

E. B. W. **Eden B Watts**
Telegraph Road, Severn
25 - Sc - 25mm("A.M."rev)
100 - Sc - 28mm("100" on both sides)

F. W. **Frank Xavier Wimmer**,1868 - 1948
Oakwood Road, Glen Burnie
(Came to Maryland from Germany in 1889. Built farmhouse in 1910 on "Gray's Luck." Checks were also used by his son, F.X.W.,Jr. and son-in-law, Frank A. Praley. FXW, Jr. farmed on the site of North Arundel General Hospital and served as a county commissioner)
1 - R
2 - R
5 - R
10 - R
20 - R
50 - R
100 - R

F. A. W. **Frank Aaron Wood**, 1877 - 1959
Elvaton Road, Millersville
(Checks found in yard)
1 - R - 21mm
5 - Oc - 19mm
10 - Sc - 25mm
20 - Oc - 22mm

G. W. **George Wiegand**
Pleasant Plains Farm
(Mentioned in ***Evening Capital***, May 24, 1900. Reported to have used checks)

G.W. **George Williams**, 1822 - 1890
Brooklyn
(Black farmer. Born a slave. Had two sons, Rudolphus Williams and Richard T. Williams who were both truck farmers using their own pickers' checks)

NV - R - 21mm - br
1 - Sc - 22mm - br
2 - Sc - 22mm - br
3 - Sc - 22mm - br
4 - Sc - 22mm - br
5 - Sc - 23mm - br
6 - Sc - 22mm - br
7 - Sc - 22mm - br
8 - Sc - 22mm - br
9 - Sc - 22mm - br
10 - Sc - 22mm - br
10 - R - 21mm - br
100 - Oc - 23mm - br

G. J. W. **George J. Wheat**
Severn

½ - Oc - 22mm - br
1 - Oc - 22mm - br
1 - R - 20mm - fr
5 - Oc - 22mm - br
5 - R - 20mm - fr
10 - Oc - 22mm - br
10 - R - 20mm - fr
15 - R - 20mm - fr
25 - R - 20mm - fr
50 - R - 20mm - fr
100 - R - 20mm - fr
200 - R - 28mm - br
500 - R - 28mm - br

H. C. W. **Henry Clifton Wigley & Charles Green Cannery**
Waterbury, Crownsville
(Tokens were believed used at the cannery, but their description is not available. Mr. Wigley was involved in many land investments and served as a county commissioner)

H. E. W. **Harry Emil Wagner**, 1882 - 1956
Admiral, (Fort Meade)
(Farmed with brother, Grover Cleveland Wagner, 1884 - 1923. Both used the HEW checks which were thrown down the old well)

½B - R - 23mm - fr
2½B - Sc - 22mm - pr
5 - R - 23mm - fr
10 - Sc - 22mm - pr

H.T.W. **Herbert Williams**, 1850 - 1933
(Reported to have used checks)

J.W. **James Franklin Wade**, 1859 - 1921
New Cut Road, Severn
(Checks also used by son Henry Randolph Wade, 1882 - 1935. Farm was purchased in the 1870s from the Curtis Creek Mining Co. and was operated by the husband of granddaughter, Ruth Wade, John Andrew Schillinger, Sr., from 1937 until his death in 1995. Farm known as Papa John's is still operated by the Schillinger family. They operate one of the few remaining roadside produce stands in Anne Arundel County)
1 - R - 20mm - fr
5 - R - 23mm - fr
5 - Oc - 19mm - pr
10 - Oc - 19mm - pr
15 - Oc - 19mm - pr
25 - Sc - 22mm - br
50 - Oc - 22mm - br
100 - Sc - 25mm - br

J.W. **John Clarence Wood**, 1878 - 1938
Elvaton Road, Severn
(J. C. Wood was killed by lightning)
1 - Sc - 25mm
2 - Sc - 25mm
5 - Sc - 31mm
10 - Sc - 31mm
$1.00 - Sc - 26mm

J.A.W. **John August Wagner**, 1849 - 1913
Telegraph Road, Admiral (Fort Meade)
(Farm was sold to Basil S. Cromwell and his brother Emory L. Cromwell. Checks were reportedly thrown down an old well)
1½ - R - 19mm - br(2var)
1½ - Oc - 23mm - br(holed)
15 - Sc - 25mm - br(2var)
15 - Sc - 22mm - br
20 - R - 24mm - rr(holed)
20 - Oc - 19mm - fr
75 - R - 23mm - fr
150 - Oc - 22mm - br
500 - R - 29mm - br(2var)

J.B.W. **Jonathan. B. Wells**
(Mentioned in ***Evening Capital***, June 7, 1900)

J.C.W. **John Clement Wood**, 1858 - 1953
Elvaton Road, Millersville
(Mr. Wood was called by his nickname,"Bud")
1 - Oc - 19mm
5 - R - 21mm
10 - Sc - 23mm
20 - Oc - 23mm

J.H.W. **James Henry Wood**, 1887 - 1953
(Family recalls checks)

J.M.W. **John Mathus Wood**
(Family recalls checks)

J. P. W. **James Pitts Watts**, 1844 - 1906
Admiral (Fort Meade)
(Father of Penrhyn S. Watts)
1½ - R - 20mm - pr - rF
10 - R - 20mm - pr - bF
15 - R - 25mm - pr - rF
75 - R - 24mm - pr - bF
300 - R - 29mm - pr -rF

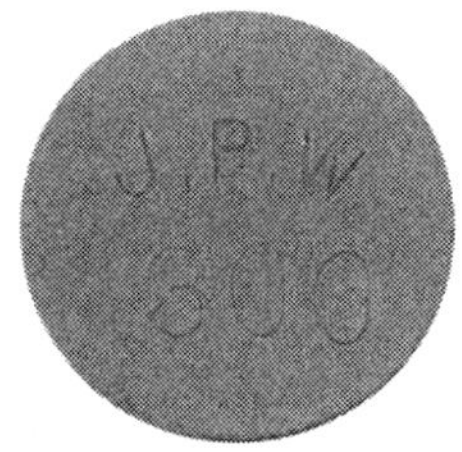

J. S. W. **James S. Wilson**, ? - 1894
Winchester
(Checks have large "J.C." on reverse. Some have "W" or "S" counterstamped. Checks were purchased by William Smith at auction)
4 - R - 20mm - fr
5 - R - 21mm - fr
10 - R - 23mm - fr
20 - R - 23mm - fr
25 - Sc - 21mm - br

J. W. W. ?
(Check found on Jenkins farm in Riviera Beach)
100 - Sc - 25mm

M. W. ?
1 - R - 24mm - pr(painted black)

P. S. W. **Penrhyn Stanley Watts**, 1890 - 1970
Admiral (Fort Meade),
(Penrhyn took over his famly farm at the age of 16, when his father, James P. Watts died. The 132 acre farm was taken by the U.S, Government when Camp Meade was established. Penrhyn purchased a new farm on Telegragh Road in 1917. The 1Qt. check was the only one used)
1Qt - R - 22mm - fr

R. W. **Rudolphus Williams**, 1856 - 1930(?)
Brooklyn
(Black farmer, son of George Williams)
8 - R - 21mm - br

R. B. W. ?
20 - R - 24mm
50 - Sc - 25mm (RBW on rev.)
100 - Sc - 25mm

R.T.W. **Richard T. Williams**, 1858 - ?
Brooklyn
(Black farmer, son of George Williams)
7½B - Sc - 23mm - br

S.W. ?
(One check found in field across from I.G.A., Route 175, Gambrills by L.W. Smith)
100(rev) - R

S.S.W. **Summerfield S. Wheeler**
Solley Road
7½ - Sc - 22mm
100 - R

S.T.W. **Samuel T. Wilson**, ? - 1945
Broadwater Beach
(Son James S. Wilson)
5 - R - 24mm - fr
10 - R - 23mm - fr
100 - Oc

T.H.W. **Theodore Hamilton Wood**, 1879 - 1960
Elvaton Road, Glen Burnie
(Son of Elijah Wood. Family recalls checks)

U.W. ?
1½ - R - br

W.W. **William Ware**, Black farmer near Ordnance Road
("B" on reverse is for Brother)
"P" - R - 20mm - fr
"B" - Sc - 22mm - br("15"rev)
NV - R - 20mm
1 - R
1½ - Sc - 25mm - br(initials on rev)
2½ - Oc - 22mm - br("50"rev)
7½ - Oc - 19mm - br
15 - Sc ("B"rev)
30 - Oc - 22mm - br
37½ - Sc - 25mm - br
75 - Sc - 22mm - br
100 - Sc - 25mm - br
500 - R - 28mm - br

W. E. W. **William Eben Watts**
Watts & Disney Roads
(Father of William Watts. Checks of 1, 6, 8, 10 also reported)
5 - R - 20mm - fr

W. H. W. **W. H. Watts**
(Some checks struck from a die)
2 – R – 24mm - rr
5 – R (die struck)
25 – R (die struck)

W. T. W. **William T. Watts**, ? - 1910
Ridge Road, Harmans
1 - Oc - 22mm - br
2 - Oc - 22mm - br
5 - Oc - 22mm - br
10 - Oc - 22mm - br
25 - Oc - 22mm - br

50 - Oc - 22mm - br
100 - Oc - 22mm - br

W. W. W. **William Wesley Wheeler, Jr.**, 1873 - 1955
New Cut Road, Severn, now part of Papa John Schillinger's farm .
(Brass checks were given to the children)

"Y"

A. K. Y. ?
NV - R - 20mm - pr
5 - R - 23mm - fr
10 - R - 23mm - fr
75 - Sc - 26mm - br
150 - Sc - 26mm - br

H. Y. **Henry Yealdhall**
Woodlawn Heights, Wellhams
(Lived at 35 Lexington St., Baltimore. Was a gentleman farmer as he practiced Law. 458 checks were donated to AACoHS)
1 - R - 23mm - br(2var)
2 - R - 23mm - br/tr(2var)
4 - R - 23mm - fr
8 - R - 23mm - fr
10 - R - 23mm - fr
100 - R - 23mm - br

"Z"

G. Z.	? 5 - R - 20mm - fr
J. Z.	**James Zeman** Obrecht Road, Millersville 5 - R - 23mm - fr
L. C. Z.	? NV - R - 24mm - fr
P. F. Z.	**Peter Frank Zeman** Elvaton, Millersville (Stepson of Frank S. Fisher. Family recalls tokens.)

Inventory By Initials Only

- A -

"Albert"
CHA
JCA
JEA
JFA
JSA & Co.
JWA
RHA

- B -

B
ALB
BBB
CB
CB Jr
C & Bro
CFB
CWB
GB
GF Beck
GHB
GMB
HB
HCB
JB
JB
JB
JFB
JMB
JRB Jr.
JSB
LMB
MB
MB
MAB
PH & B (Bro.)
RTB
TB
TFB
UWB
WB
WLB
WOB
WPB

- C -

C & Bro.
BFC
BFC
BSC
CC
CHC
CWC
EC
EDC
ELC
FC
FPC
FWC
GTC
GWC
HC
HAC
HDC
HMC
HWC
JC
JHC
JHC
JMC
JSC
JTC
JWC
MC
PC
PAC
RNC
RSC
RWC
SAC
SSC
STC Jr.
STC of J
TAC
TMC
TWC
VBC
WC
WHC
WHC
WLC
WSC
WTC

- D -

D
Dickler
D Bros.
AJD
CMD
DFD
GD
HD
JD
JGD
JTD
MD
RD
SED
TD
UGD
WD
WD
WPD
WTD

Inventory By Initials Only - continued

- E -

GWE
JFE
OE
WE

- F -

DF
FSF
GFF
HF
JF
JGF
JHF
HCF
RTF
WF

- G -

BG (2)
CG
CBG
CMG
FAG
GAG
HG
HEG
HEG
JHG
KG
MG
PXG
RWG
SLG
TJG
WG
WMG
WTG

- H -

AH
ASH
BH
CAH
C & RTH
CHH
CHH
CH
DLH
EHH
EJH
EJH Jr.
FLH
HH
HRH
JH
JH
JH
JAH
JBH
JGH
JHH
JLH
JSH
JTH
MH
OH
PH
PH & B
RWH
RWH
SPH
WSH
WTH
WWH

- J -

CNJ
EJ
EGJ
FJ
FJ
GAJ
GCJ
GGJ
JAJ
JCJ
JHJ
JIJ
SJ
TWJ
WEJ
WTJ
ZJ

- K -

K
BAK
CFHK
EK
EJK
HFK
HNK
NSK
JK
JHK
LK
LCK
TRK
WDK

Inventory By Initials Only - continued

- L -

AL
ASL
CCL
FL
J Libaros
JML
MJL
SL
SL Jr.
WL
WJL

- M -

McK
AM
AGM
BFM
FCM
FM F. MEWSHAW
GMc/GMcP
GM
GGM
GMM
HM
HHM
HJM
JM
JSM
LM
RHM
WJM
WRM

- N -

N
CAN
JE Nunn
J & N
RGN
RMN

- O -

HO
JO
REO
TO

- P -

P
P Jr
ACP
BFP
CP
CAP
CBP
CGP
ELP
FP
JP
JDP
JEP
HP
JRP
KP
LBP
NP
NAP
NCP
OCP
OSP
RDP
RHP
RLP
RNP
STP
TP
TCP
TOP
WP
WJP
WJP
WMP
WWP

- R -

R
BTR
EWWR
FSR
JR
JBR
JHR
JSR
JWR
RR
WR/W Remmey
WTR

- S -

AS
AAS
AAS
ACS
BHS
CHS
CLS
CWS
CVB Smith
D. St
ES
ES
EAS
ELS
FS
FMS
FOS
GS
GBS
BCS
GM & S
HS
HBS
HJS

Inventory By Initials Only - continued

HLS
HS
JCS
JES
JFS
JS
JS
JWS
LJS
L. SCHR
LRS
MS
MFS
NS
NJS
PLS
RAS
RLS
ROS
ROS
THS
TSS
WS
WAS
WES
WJS

- T -

T
BFT
CRT
D. St
JFT
LCT
LJT
OT
SST
TJT
ZT

- U -

JWU
TBU

- V -

V
JV

- W -

ACW
CW
CRW
EW
EW
EBW
FW
FAW
GJW
GW
GW
HEW
HTW
JW
JAW
JBW
JCW
JHW
JMW
JPW
JSW
JWW
MW
PSW
RBW
RTW
RW
SW
SSW
STW
THW
UW
WW
WEW
WHW
WTW
WWW

- Y -

AKY
HY

- Z -

GZ
JZ
LCZ
PFZ

Index

A

B

C

D

E

F

G

H

I

J

K

L

M

N

O

P

Q

R

S

Patron List

The publication of *Strawberries, Peas & Beans* has been made possible through the generosity of the following Patrons:

Val Alexander & Helen Herndon
The "Val" uable "Hel" p Team of Long & Foster Realtors

Mr. & Mrs. Tyras S. Athey

Robert C. Baldwin

The Bank of Glen Burnie

James M. Boswell

Bunker Hill Farms

Ralph & Mary L. Cadogan

Frank & Pat Calvert
& Family

James B. & Mary M. Calvert

Katherine Mumford Cave

Thelma A. Cole

Jeff Cook

Edward L. Craig

William S. Crisp

Patron List *(continued)*

Isabel Shipley Cunningham

Roy F. & Rosemary B. Dodd

Marie Angel Durner

Gable Signs & Graphics, Inc.

Nancy C. Gaston
In Honor of
Helen Cole Gardner

Mr. & Mrs. Leslie E. Goldsborough, Jr.

Harry D. Greenwell

Louis G. Greenwell

Carroll & Carolyn E. Griffith

Jean Boschert Guite

Millard W. Hajek

Henry & Carolyn Hein

Mary Hoffman

Forrest G. & Mildred B. Hoffmaster

Charles & Inge Hofmeister

Dave & Pam Hovatter

Patron List *(continued)*

Thomas Ingram

Jack & Wallis Kelbaugh

Kirkley-Ruddick
Funeral Home

Cathie Kline

F. William & Beverly R.
Kuethe, Jr.

Kurtz's Beach, Ltd.

Edwin A. Lechowicz

Alvin W. & Madeline Lenz

O. James Lighthizer

Alfred J. & Irene H. Lipin

Amy E. Mahanes

Albert Marecki

Ted & Ellen Mathison

Charles R. Maynard

Barrett & Anne McKown

William & Carolyn Melton
MARCOM Marketing &
Communications

Patron List *(continued)*

Mr. & Mrs. Gilbert H. Moore, Jr.

James R. Morrison

David Royal Mumford
& Family

Elaine Mumford

Laura Mumford &
Penny Davis

Stanley W. & Anne C. Myllo

Dr. & Mrs. Robert W. Neidlinger

Louise Benson Nichols

Odenton Heritage Society

Jo Shipley Owens

Pumphrey Farm

Tom Redmond

Orlando Ridout IV

Raymond & Elsie (Remmey)
Ringgold

B. A. Robison, Jr.

Rose Family Partnership

Patron List *(continued)*

Margaret Watts Rose

George & Kathy Sauble

Mark, Ann, Barbara
& Stephen Schatz

Emma M. Schramm

William M. Schramm

Sherblom - Woodward Family

James W. Shipley

Lois Harman Simpson

Dorothy M. Singleton

Amy Mumford Sirotek

Susan Snyder

Joel Sparhawk
Severn Graphics, Inc.

June Sappington Spiess

Hilary L. Stallings, Sr.

Jeanne A. Stinchcomb

Sara Anne Stinchcomb
& Family

Patron List *(continued)*

Paula & Frank Tremel

William & Lois Upton

Edward C. Vernon

George F. Ward

George F. Ward, Jr.

Richard K. & Elizabeth M. Wengert, Sr. & Children

Stanley F. & Elizabeth Kelbaugh Westendorf

Mary Lou Wilcox

John F. & Barbara B. Wing

Nancy L. Zastrow

John H. Smith

in Memory of His Beloved Wife

Ora Pumphrey Smith

To the Greatest Dad
and Granddaddy

We congratulate you on
Realizing Your Dream!

All our love,

Amy, Rob & Amanda Sirotek

CARING COVE